SCIENCE AND RELIGION

Other interview books from Automatic Press ◆ $\frac{V}{1}$ P

Formal Philosophy
edited by Vincent F. Hendricks & John Symons November 2005

Masses of Formal Philosophy
edited by Vincent F. Hendricks & John Symons October 2006

Philosophy of Technology: 5 Questions
edited by Jan-Kyrre Berg Olsen & Evan Selinger February 2007

Game Theory: 5 Questions
edited by Vincent F. Hendricks & Pelle Guldborg Hansen April 2007

Philosophy of Mathematics: 5 Questions
edited by Vincent F. Hendricks & Hannes Leitgeb January 2008

Epistemology: 5 Questions
edited by Vincent F. Hendricks & Duncan Pritchard September 2008

Philosophy of Medicine: 5 Questions
edited by J.K.B.O. Friis, P. Rossel & M.S. Norup September 2011

Narrative Theories and Poetics: 5 Questions
edited by Peer F. Bundbaard, Henrik Skov Nielsen & Frederik Stjernfelt 2012

Philosophical Practice: 5 Questions
edited by Jeanette Bresson Ladegaard Knox &

Jan Kyrre Berg Olsen Friis January 2013

Intellectual History: 5 Questions
edited by Morten Haugaard Jeppesen, Frederik Stjernfelt & Mikkel Thorup
May 2013

Philosophy of Nursing: 5 Questions
edited by Anette Forss, Christine Ceci & John S. Drummod
October 2013

See all published and forthcoming books in the 5 Questions series at
www.vince-inc.com

SCIENCE AND RELIGION: 5 QUESTIONS

EDITED BY

GREGG D. CARUSO

Automatic Press ♦ ⊻⊤P

Automatic Press ♦ $\frac{V}{I}$ P

Information on this title: ww.vince-inc.com

© Automatic Press / VIP 2014

First published 2014

Printed in the United States of America
and the United Kingdom

ISBN-10 / 87-92130-51-8
ISBN-13 / 978-87-92130-51-8

Cover design by
Vincent F. Hendricks

Contents

ii

Preface

––––––––––––––––––––– ♦ –––––––––––––––––––––

The relationship between science and religion has been a subject of profound interest for philosophers, scientists, and theologians for centuries. This should be of no surprise given that the scientific and religious worldviews dominate our understanding of the cosmos and our place in it. It is important, therefore, that we ask ourselves: Are science and religion compatible when it comes to understanding *cosmology* (the origin of the universe), *biology* (the origin of life and of the human species), ethics, and the *human mind* (minds, brains, souls, and free will)? Do science and religion occupy non-overlapping magisteria? Is Intelligent Design a scientific theory? How do the various faith traditions view the relationship between science and religion? What, if any, are the limits of scientific explanation? What are the most important open questions, problems, or challenges confronting the relationship between science and religion, and what are the prospects for progress? These and other questions are explored in *Science and Religion: 5 Questions* — a collection of thirty-three interviews based on five questions presented to some of the world's most influential and prominent philosophers, scientists, theologians, apologists, and atheists.

Contributors include a Nobel Prize winning physicist, three Templeton Prize winners, two "Humanist of the Year" winners, the "Most Influential Rabbi in America" (*Newsweek*, 2012), "the leading American expert on Tibetan Buddhism" (*New York Times*), a National Humanities Medal winner, a National Medal of Science winner, a Star of South Africa Medal winner, a Carl Sagan Award winner, a National Science Board's Public Service Medal winner, a MacArthur Fellow, a Lakatos Award winner, an Erasmus Prize winner, a "Friend of Darwin Award" winner, a "Distinguished Skeptic Award" winner, the first Muslim to deliver the prestigious Gifford Lectures, and many more.

This collection brings together a diversity of different perspectives and traditions. It does not attempt to promote or argue for any particular point of view — it leaves it to the reader to draw his/her own conclusions. And while, of course, I have my own views on these matters (as any thoughtful person probably should), as editor I have made a concerted effort to put together a balanced collection. In the following you will find representatives of the three major theistic religions (Christianity, Judaism, and Islam) along with Buddhism, as well as some of the world's leading atheists.

Given the diversity of perspectives represented here, as well as the accessibility of the interviews, this collection will serve as an excellent introduction for anyone interested in the intersection between science and religion. It is perfect for general audiences as well as for courses in Science and Religion, Apologetics, and Philosophy of Religion. It should also be of interest to professional philosophers, scientists, theologians, and apologists looking for a concise "snapshot" of what some of the leading figures in their respective fields currently think about these important issues.

Finally, a note on how the interviews were conducted: All contributors were presented with the same five questions, with their responses appearing here in full. The majority of interviews were conducted through electronic correspondence, which allowed contributors ample time to develop their responses. A few interviews, however, were conducted via Skype and later transcribed. While contributors approved transcriptions, any remaining errors are the sole responsibility of the editor.

Corning, NY, March 2014
Gregg D. Caruso

Acknowledgements

I would like to thank Automatic Press ♦ $\frac{V}{I}$ P, especially editor-in-chief Vincent F. Hendricks and associate editor Henrik Boensvang, for their enthusiastic support of this project and their assistance in bringing it to fruition. I would also like to thank Katherine Douglas, Marian Eberly, and Byron Shaw for encouraging my research endeavors, including my editorial work on this and other volumes. A special debt of gratitude is owed to my wife and daughter for their eternal wellspring of love and support—I simply owe you everything! I would also like to thank the following friends, family members, and colleagues: Louis and Dolores Caruso, George and Marina Kokkinos, Robert Talisse, David M. Rosenthal, Bruce Waller, John Greenwood, Michael Levin, Neil Levy, Owen Flanagan, Josh Weisberg, Thomas W. Clark, Steven Pinker, Jerry Coyne, Noam Chomsky, Philip Clayton, Thomas Nadelhoffer, Jesse and Feighanne Hathaway, Andrew Hathaway, James and Melissa Cruz, John Celentano and Shelly Bhushan, Mark McEvoy, Rick Repetti, Sandy Turner-Vicioso, Michael Beykirch, Mary Guzzy, Vince Lisella, Louis and Angela Caruso, Larry and Mary Caruso, Robert Caruso, Darlene Caruso, Peter and Marian Caruso, Thomas Caruso, Beth Caruso, Asterios Kokkinos and Justine Barron, Costa Kokkinos and Miriah Harvey, and Joey Kokkinos and Emily Shin. Finally, I am especially grateful to the contributors of this volume for agreeing to participate and for making this project not a mere possibility, but a reality.

Corning, NY, March 2014
Gregg D. Caruso
Editor

Introduction

—————————— ♦ ——————————

In 1997, the late paleontologist and evolutionary biologist Stephen Jay Gould famously proposed "a blessedly simple and entirely conventional resolution to...the supposed conflict between science and religion." His proposal is now famously known as the thesis of *non-overlapping magisteria*—or the *NOMA thesis*.[1] A *magisterium* refers to a domain of teaching authority. And the thesis of non-overlapping magisteria (or NOMA) maintains that "the magisterium of science covers the empirical realm: what the Universe is made of (fact) and why does it work in this way (theory). The magisterium of religion extends over questions of ultimate meaning and moral values." According to Gould, since these two magisteria do not overlap there is no real conflict (or at least there shouldn't be) between science and religion. As Gould put it, "science studies how the heavens go, religion studies how to go to heaven."

While Gould's attempt at a peaceful resolution between science and religion satisfied some, many were not convinced. As Richard Dawkins, the world-renowned evolutionary biologist and "new atheist," argued:

> [I]t is completely unrealistic to claim, as Gould and many others do, that religion keeps itself away from science's turf, restricting itself to morals and values. A universe with a supernatural presence would be a fundamentally and qualitatively different kind of universe from one without. The difference is, inescapably, a scientific difference. Religions make existence claims, and this means scientific claims.[2]

Who, then, is correct? Do science and religion occupy non-overlapping magisteria? Furthermore, do science and religion *each* have a magisterium (domain of teaching authority) to speak of? And if so, do they stay within their respective domains? This is just one of the five questions posed to the thirty-three contributors of this volume.

Another question (perhaps the key question) presented to the contrib-

---

[1] See Gould's essay "Non-Overlapping Magisteria," *Natural History* 106 (March 1997): 16-22, and his subsequent book *Rocks of Ages: Science and Religion in the Fullness of Life* (1999).

[2] Richard Dawkins, "Obscurantism to the Rescue," *Quarterly Review of Biology* 72 (1997).

utors goes directly to the heart of the matter: Are science and religion compatible when it comes to understanding *cosmology* (the origin of the universe), *biology* (the origin of life and of the human species), *ethics, and the human mind* (minds, brains, souls, and free will)? Certainly the relationship between science and religion has been, at least at various points in history, a rocky one. Just ask Galileo Galilei or Giordano Bruno, both of whom were persecuted by the church for their scientific teachings. Or consider the debate in the United States over evolution—e.g., the 1925 *Scopes Trial* or the more recent 2005 *Kitzmiller v. Dover Area School District* case in Harrisburg, Pennsylvania. Of course, these examples do not mean that there is an essential or necessary conflict between science and religion. Perhaps these historical cases are better explained by pointing to misinterpretations or misunderstandings, or the overstepping of one's *magisterium*, or political causes rather than religious, etc. Nor does it mean that all religious traditions are created equal. Some religious traditions may be more or less compatible with our best scientific understanding of the world. Nonetheless, most religions (with perhaps the exception of Buddhism) have origin stories, cosmogonies about the origin of the universe and claims about the origin of human life. They also maintain certain metaphysical claims about the nature of the self, including claims about the existence of souls, life after death, reincarnation, and/or free will. The question, then, is whether these religious accounts of creation, the nature of the human mind, and our place in the universe, can be reconciled with our best scientific accounts of the same?

Ultimately, I leave it to each contributor to define what he or she means by *science* and *religion* and to judge the compatibility or incompatibility of the two. Some of the contributors characterize the relationship as one of conflict, others describe it as one of harmony, and others (still) propose that there is little interaction between the two. Readers must decide for themselves, but I can think of no better guides than the ones assembled here. Given the caliber of contributors involved, and their varied contributions to theorizing about science and religion, the interviews contained herein provide an important resource for those seeking their own answers as well as a useful snapshot of the current state of play in the ongoing debate between science and religion.

The remaining questions provide additional insight into the lives and work of the contributors, exploring their most important contributions and how they first became interested in the intersection between science and religion, as well as a consideration of the most important *open* questions, problems, and challenges confronting the relationship between science and religion and the prospects for progress.

1

Simon Blackburn

Simon Blackburn is an internationally recognized British philosopher known for his work in philosophy of mind, philosophy of language, and philosophy of psychology, as well as his efforts to popularize philosophy. He retired as Professor of Philosophy at the University of Cambridge in 2011, but remains a Distinguished Research Professor in Philosophy at the University of North Carolina at Chapel Hill, a Fellow of Trinity College, Cambridge, and a member of the professoriate at the New College of the Humanities. Blackburn is the author of hundreds of scholarly articles and over a dozen books, including *Spreading the Word* (1984), *Essays in Quasi-Realism* (1993), *The Oxford Dictionary of Philosophy* (1994), *Ruling Passions* (1998), *Think* (1999), *Being Good* (2001), *Lust* (2004), *Truth: A Guide for the Perplexed* (2005), *Plato's Republic* (2006), *How to Read Hume* (2008), *The Big Questions: Philosophy* (2010), and *Practical Tortoise Raising and other Philosophical Essays* (2010).

1. What initially drew you to theorizing about science and religion?

I came from a standard English secular background, in which none of my family were religious by any stretch of the imagination, but in which the Church of England was regarded with benign toleration. It kept out of our way, and by and large we reciprocated. Clergymen were often figures of fun, which could nevertheless alternate with respect when the occasion demanded it. For instance, a christening, wedding, or funeral would see us in church, and it would be important to show due decorum. And one would never dream of interrogating a clergyman about the "basis" of his faith (there were no women clergy in the Church of England in those days). Nobody cared about doctrine, although Protestants were clearly better than Roman Catholics, partly no doubt because the Pope was always a foreigner. But this too was not particularly important. Because it had a good reputation some friends of my family sent their daughter, a bright little girl, to a primary school run by nuns. When one of the nuns asked "Do you little girls know what religion we are?" she shot up her hand and gave the answer " Yes. You are what my daddy calls Bloody Catholics." He had to go and apologize, although I can't imagine what he found to say.

I do not think I had any doctrinal views of my own, although I do remember one philosophical view. In order to encourage us to shut our eyes piously during the perfunctory prayer which started the day we were told by some teacher that if all our eyes were shut angels would manifest themselves, although however suddenly or unpredictably we peeked, they would already be gone. I remember puzzling over the epistemology of this for a while, and later when I began to worry about quantum theory and the fact of observations interfering with the phenomena at which they are directed, I felt quite familiar with the problem.

Religious observance thrust itself more into my life when I went away to school, at the age of twelve. Like most public schools at the time, mine took a backbone of Christian observance for granted, with Chapel to start the day, and prayers to finish it. I did not mind this for a while, since the liturgical language and the music both had their appeal. In any case, children take a lot of the odd things adults insist on for granted. But as a rather rebellious adolescence kicked in the whole thing began to lose its grip. I think my disenchantment was aided by my disliking of one of the chaplains, who struck me as both complacent and vain, and since these were clearly qualities inconsistent with Christian humility, undoubtedly a hypocrite as well. It is probably an exaggeration to say that this was a loss of faith, since I don't suppose I ever really had any. I certainly suffered no dark nights of the soul, tussles with Christ, existential agonies, fear of hell and the like. It was just rather thankfully leaving aside a jacket that had never really fitted.

I was on the science side at school (in the UK, we had at the time to specialize after the general education examination, which I took rather young). But mathematics, physics, and chemistry had little or nothing to do with my rejection of the words and rituals that we were supposed to be so solemn about. I don't think it ever occurred to me that science provided some kind of additional argument that these words and rituals needed to be abandoned, and in this, as I shall later try to explain, I think I showed quite good taste. I thought common sense, or what I might now classify as everyday experience, was quite enough to make it unlikely that anyone had ever lived for nine hundred years, turned water into wine, risen from the dead or gone off like a human rocket to some haven beyond the sky. I thought that the most casual observation of the world conclusively showed that it was not the product of an all-knowing, all-powerful, all-benevolent creating and sustaining hand. Suffering did not need science to detect it.

I think the first book I read which chimed in with my views at the time was Bertrand Russell's *Why I am not a Christian*, which gave me an excellent store of ammunition whenever religious discussions took place. Later, when I went to Cambridge, and discovered much more about the philosophy of science, and in particular the empiricist departure from rationalism, Hume began to supplant Russell as my favourite infidel.

2. Do you think science and religion are compatible when it comes to understanding cosmology (the origin of the universe), biology (the origin of life and of the human species), ethics, and/or the human mind (minds, brains, souls, and free will)?

I do think they are compatible, but perhaps for a rather unusual reason. This is because I do not think religion at its core or at its best purports to deliver any understanding about these things, or indeed about anything. I think it serves other functions, for good or ill. Of course, particular religionists, meaning by this term those who suppose themselves to be following some faith and appraised of some doctrines, may hold views that are absurdly at variance with those of science. Faced with some fundamentalists, of whatever faith, we might go down a checklist, and find that they hold various doctrines quite inconsistent with those of science: that the earth is around 6000 years old, that the biblical flood floated Noah's ark far up Mt. Ararat, that homo sapiens is no relation of other primates, that the sun goes round the earth, that polio vaccines are lethal, or that the 1997 Hale-Bopp comet was in fact a taxi for dead Californians to go to a better place. That is a pity. But I am sure many theologians would join me in denying that those aberrations are any part of religion at its core or at its best.

Of course, that leaves the question of what is the essential core that remains, and this will be highly contestable. If the essential core of Christianity, for instance, is supposed to contain historical commitment to miraculous events, and if miraculous events are inconsistent with our best-attested scientific laws, then once more there is a straightforward clash between religion and science. But if it is allowed that we can read the sacred texts differently, not as history but as myth, metaphor, or poetry, or simply as a source of moral examplars, then again the clash can be avoided.

The list in this question includes cosmology and biology as actual sciences, and brains as a very definite topic for science. If religionists include armchair or dogmatic opinions about these in their doctrines then of course they risk clashing with science and will probably do so. But the list also includes minds, souls, and free-will. Science can certainly need taking into account when we talk about these. But it may not be nearly so clear whether the things that scientists say about them are backed by the authority of scientific investigation, or whether they involve philosophical assumptions and perhaps mistakes. For example, a neurophysiologist might deny the existence of mind, having interpreted it as a kind of ghostly entity distinct from yet somehow running parallel with the events he discovers in the brain. Or, to take a recent example, he might deny the reality of free-will because he discovers "readiness potentials" in the motor cortex prior to the apparent moment of conscious decision. Both of those interpretations of the science would involve philosophical claims (and mistakes) about the nature of mind or free-will. Insofar as religions also make those mistakes, it is philosophy, not science, that confronts them with its flaming sword. But even within Christianity different sects have held wildly different views about minds, and especially about free-will.

I now want to explain what I mean by religion "at its core or at its best." I do not think there is a better definition of religion than that given by Émile Durkheim a century ago:

> A religion is a unified system of beliefs and practices relative to sacred things, that is to say things set apart and surrounded by prohibitions—beliefs and practices that unite its adherents in a single moral community called a church.

The practices may include ritual, music, drama, song, dance, poetry, myth, building, and painting. They may have other functions than unifying the congregation: for instance, they may offer consolations and hopes to those who are afflicted, or they may cement sympathies and fellow-feelings within the group, and otherwise adjust their moralities and their politics. They may encourage awe at the wonder of the cosmos, amazement that the laws of nature exist, or that the various fine-tuned cosmological constants on which it all depends have just the right values to keep the whole show going. But in order to do these things the rest of the practices may completely eclipse the belief part, and in many religions they do. Indeed, they can occupy the whole space: Buddhism, Christianity, Hinduism and Islam each have "apophatic" traditions which hold that the sacred is best approached by silence rather than doctrine, although it is a silence that has to spill over into the adept's way of being in the world. In a nice simile I read recently, these traditions hold that religions are like public swimming pools, in which the most noise comes from the shallow end. Unfortunately the shallow end is one in which both religionists and their opponents are likely to be found splashing about.

Insofar as apophatic traditions interpret the core or the best part of religion, obviously there is no clash between religion and science, because religion is not interested in advancing empirical or theoretical doctrines at all. It may ask for an emotional response to the world, but an emotion like awe or wonder cannot be foreign to science, at its best. Of course, whether the church or congregation into which a religious practice has welded people is admirable or not remains subject to moral appraisal. The animating spirit may be benign, tolerant, humane, inclusive, and peaceable. Or, it may be malevolent, fearful, intolerant, inhumane, divisive, and warlike. The madness of crowds is a dangerous thing, and one practice that unites people very efficiently is hatred of those outside the community. If a congregation partakes more of the second menu than the first it presents a moral problem, but not because of its scientific ignorance.

3. Some theorists maintain that science and religion occupy non-overlapping magisteria—i.e., that science and religion each have a legitimate magisterium, or domain of teaching authority, and these two domains do not overlap. Do you agree?

I think this is a silly idea, for reasons that are probably evident from my previous answer. I do not think that religion has any magisterium, or area of knowledge of its own. On the contrary, to perform Durkheim's function of uniting a group into a congregation there has to be an element of arbitrariness, and an element of the mysterious. An emotional charge has to be attached to something: a particular place, a mode of dress, a set of symbols, a practice available to those on the inside and not to those outside. These things will be taught to initiates, so in that thin sense we might say that there is a "magisterium": we are the people who do not climb the sacred mountain, or eat shellfish or cows, or we are the ones who perform female genital mutilation, or we are the knights who say "ni." But this is not instruction in the way of the world or the way of the universe. It is just instruction in the way of the tribe.

Again, this is concentrating on the core of the matter. At the periphery the priests or wise men or women of a tribe may have doctrines to impart, about medicine, or health, or their environment, and these may be true or false. They will be subject to empirical test, like any other such claim. It may sometimes be wise for science to take notice of them. Similarly a religious tradition might also contain words of wisdom about morality: stoicism, for instance, straddles the line between religion and philosophy. We would be foolish to jettison the thought of an Augustine or an Aquinas wholesale, for they may have things to teach us about human beings and human life.

4. What do you consider to be your own most important contribution(s) to theorizing about science and religion?

I do not claim to have made any important contributions to theorizing about either science or religion. All I may claim to have done is to emphasize doctrines of both Hume and Durkheim that seem to me to have been forgotten in the current, rather primitive, debate between "new atheists" and religionists.

This is, perhaps, worth explaining. Some two hundred and fifty years ago David Hume showed that if our thoughts fly up to a supernatural creator when we look at the world around us, the implications must be less than meets the eye. In fact, there aren't any. For even if we convince ourselves of a First Cause or a Divine Architect, then all it implies is that he (or she or it or they) made a world like this. It is our only data point. If the unjust flourish in this world, then that is how it is. If the innocent die or suffer, while the evil prosper, that is how it is too. That's what this architect does, and *for all we can possibly tell*, it is all he ever does. Anything else in which we dress him (or her or it or they) will be the result of our own fancies. Our fancies may be important to us: we can convince ourselves that we saw the statue move, that the miracle took place, that Divine favour is ours, that the voice in our head belongs to Him, and a thousand such things. These may move us to be good or to be bad. But they are our own stuffing. They are not implied by any doctrine for which there can be the shadow of a reason.

Once we understand this, it does not much matter whether we argue for the Divine Architect or not. No trip to Hotel Supernatural enables you to come back with more legitimate luggage than you took there. Legitimate luggage is our own best science or understanding of the way of the world as it lies open to our sense experience. It also includes our own best understandings of virtue and vice, duty and obligation, law and the political order, and a secular understandings of those things, whether or not it builds on whatever was best in a religious tradition, can be very good indeed. Illegitimate luggage is only the weight of other peoples' dogmas, or the uncritical reception of the views of some particular tribe of thousands of years ago.

When I call the debate between new atheists and religionists primitive, I am not simply being rude to both sides. I have cheerfully admitted to being an infidel, but I do not regard that as equivalent to being an atheist. An atheist finds an ontological claim that he takes the religionist to be making, and finds that claim intelligible enough to understand, but then wants to deny it. The agnostic agrees that there is such a claim, but then says he suspends judgment on it. I do not think this is ground worth fighting over. If a religion at its core or at its best takes the form I have described, there is no intelligible ontological claim to be asserted or denied or about which to be undecided. The words that are to be said, and said with all due solemnity, may take the superficial form of an ontological claim: "I know that my redeemer liveth," or "God's eye is all-seeing." But their function is quite different: it is to orientate the adept towards the world in a way that is a part of the way of life of the congregation or community. It is, as it were, part of a socially binding yoga. To perform this yoga, myth is as good as history, and cloudy imaginings may be as good or better than intelligible doctrines.

Hume never called himself an atheist, precisely because he saw that this is not where the important issues lie. Those issues lie in the moral and political implications that people take themselves to be able to draw from their supernatural imaginings. Since there can properly be no such implications, the imaginings can be left to themselves. Hume's targets were the self-deceived, the charlatans, and the enthusiasts who rush into the void with their own fantasies and dogmas, and the frailties of the human mind that enable them to get an audience.

5. What are the most important open questions, problems, or challenges confronting the relationship between science and religion, and what are the prospects for progress?

The scientist ought to thank Hume for inaugurating the natural history of religion, or in other words the comparative, empirical understanding of religious practice as a human phenomenon. But in turn that might require a retraction by some of the more vociferous cheerleaders for science in these debates. The religionists they oppose can no longer be seen simply as particularly stupid and ignorant, superstitious, benighted discredits to the human race. They have an emotional investment in a practice that has survived because of its important social and moral functions. The scientist perhaps has no such investment. He may pride himself that this is a matter of his great intelligence, but it may be just as much a matter of his or her social context. Nearly all the early, glowing, exemplars of the scientific spirit in action in the West, including Copernicus, Descartes, Leibniz, Galileo, Newton, Boyle and countless others were good Christians.

However the standpoint I have sketched asks more concessions from the usual religionist than from any scientist. Religionists may have to relinquish their own understanding of their practice. The refined religionist, indeed, may already be happy in the apophatic understanding of this. But the unrefined may have thought themselves first to be following a chain of argument up into the supernatural world, and then following a chain of implications back from that to serve as their authoritative, established signpost in this world. They may take the historic occurrence of miracles to be established, and to accredit the objects of their devotions with supernatural privileges. I do not doubt that it may be painful, and even impossible to realize that this is deceptive clothing, but that the real body of the practice is quite different. The emotional and social needs that the practice serves stand as a gigantic obstacle to accepting this understanding.

I doubt if there are any prospects for progress. Partly this is because I think the accelerating and indefensible allocation of benefits and privileges in the winner-take-all capitalist world we are in, encourages religion in those outside the richest strata. Having next to no goods in this world they can only salvage their identities and pride or only find consolation and hope in the promises of a mysterious salvation. Militant atheists often like to remember Marx's dictum that religion is the opium of the people. They are less likely to remember the context of his remark:

> Religious suffering is, at one and the same time, the expression of real suffering and a protest against real suffering. Religion is the sigh of the oppressed creature, the heart of a heartless world, and the soul of soulless conditions. It is the opium of the people. The abolition of religion as the illusory happiness of the people is the demand for their real happiness. To call on them to give up their illusions about their condition is to call on them to give up a condition that requires illusions. The criticism of religion is, therefore, in embryo, the criticism of that vale of tears of which religion is the halo.

Hume thought that primitive religion arose from personifying terrifying natural forces, and hoping to placate them. These remarks, from his *Critique of Hegel's Philosophy of Right*, show Marx supplementing Hume with a more pointed economic and political diagnosis. Marxism was a call for people to transcend a condition that requires illusions. But it failed. It seems, then that a condition that requires illusions will always be with us, so if Marx was right, religion will be as well, and probably not often at its best, either.

2

Susan Blackmore

Susan Blackmore is a freelance English writer, lecturer, and broadcaster, and a Visiting Professor in Psychology at the University of Plymouth. She researched the paranormal for many years but is now best known for her work on consciousness, free will, evolutionary theory, and memetics. She is a Fellow of the Committee for Skeptical Inquiry (formerly CSICOP) and in 1991 was awarded the CSICOP Distinguished Skeptic Award. An atheist who has criticized religion sharply, she has also been practicing Zen for over thirty years (although she does not consider herself a Buddhist). She has authored over sixty academic articles, contributed to over eighty books, authored fourteen books of her own, and is a regular contributor to *The Guardian* newspaper and *Psychology Today*. Her books include *Parapsychology and Out-Of-Body Experiences* (1978), *Beyond the Body: An Investigation of Out-Of-Body Experiences* (1982), *Dying to Live: Near-Death Experiences* (1993), *In Search of the Light: The Adventures of a Parapsychologist* (1996), *The Meme Machine* (1999), *Consciousness: An Introduction* (2010, 2nd Ed.), *Conversations on Consciousness* (2005), and *Zen and the Art of Consciousness* (2011).

"LIKE SCIENCE, THE BUDDAH'S IDEAS CHALLENGE EVERYTHING WE LIKE TO BELIEVE"

1. What initially drew you to theorizing about science and religion?[1]

That is lost in the mists of time. What I do remember as a young child is being fascinated by big questions and those questions were both scientific and religious. Who am I? Where am I? What am I? Why am I here? What's the meaning of it all?

I have one particular memory, a sort of flashbulb memory of walking around the pond at my parents' house, when I must have been about six or seven, thinking about the nature of heat. What on earth could heat be? I remember holding my hand out in front of me and thinking, "My hand's warm and the pond is cold. What is going from one to the other?" and not knowing and thinking that it was maybe some kind of wiggling of my hands. I don't know if I had heard about heat being energy—I have no idea—but that sort of thing fascinated me as a child.

And also I had huge arguments with my mother because I had religious phases and anti-religious phases and my mother was a Christian. I would tell her, "Yeah, but there couldn't be a God because if he did

---

[1] This interview was conducted via Skype and the transcription approved by Susan Blackmore. Any remaining errors or mistakes are the fault of the editor.

exist, why does he do bad things?" and all those classic arguments.

I suppose in a way the answer is that I've always wondered about science and religion, but then as a student in 1970 I had a most dramatic experience that changed my life, and I mean that quite sincerely. Almost all the research I've ever done since then has depended in some way upon that experience. I have written about it in at least two of my books and online, and I still don't really know what to call it.

In the beginning I called it "astral projection." Then I learned the phrase "out-of-body experience." Then when the term "near-death experience" was invented, which was in the mid-70s, I called it a near-death experience. In fact I wasn't actually near death, and yet my experience included almost all the features of a classic near-death experience. But I could also call it an "exceptional human experience," which is Rhea White's term, or I could call it a "mystical experience," which it certainly was. The latter phases of it had many of the features of classic mystical experiences.

That experience, in which I seemed to be out of my body observing somewhere else, made me absolutely sure that I had a soul or a spirit and that my soul or my astral body had left my physical body and gone to other realms. Yet I was at Oxford University studying physiology and psychology. So this horrendous clash between what I knew from my own experience and what I was learning in my science absolutely electrified my mind. It was wonderful. I was really enjoying my degree and yet this clash made me absolutely determined to prove to the world that there really is a soul or a spirit; that there really is life after death; there really is telepathy, clairvoyance, and all of those paranormal things, which seemed to me to be obviously true based on my own experience.

That sort of confidence lasted a few days or a few weeks, but even during the experience itself my skeptical nature was creeping out and asking, "Yes, but is this really so?" "Are those roofs I'm seeing really the roofs and the gutters around this building?" and the next morning I looked, and they weren't right and so my skepticism started to creep in. And to cut a long story short, over the years I went into research in the paranormal, did a Ph.D. in parapsychology, did lots of research and finally became very, very skeptical to the point that I am as sure as I can be—which is not 100%—that there are no paranormal phenomena, that there is no soul or spirit, and there is no personal life after death.

That experience really underlies, I would say, all my subsequent theorizing about science and religion. In a way, that's the answer to your question.

2. Do you think science and religion are compatible when it comes to understanding cosmology (the origin of the universe), biology (the origin of life and of the human species), ethics, and/or the human mind (minds, brains, souls, and free will)?

No, but I want to qualify that. First of all, I'd like to go through each of these topics in turn because I think the answer is different for each of them. But also, I think we need to be clear what we mean by "religion." Do we mean the great monotheistic religions? Do we mean specific religions? Do we mean religious tendencies or religiosity, or do we perhaps even mean something that one might call spirituality that might not involve either gods or rituals or other features of the classic religions?

I will try to answer with respect to the religions that I know something about, because I don't know a lot about all religions, obviously. I know quite a lot about Christianity because I was brought up in the Church of England. I was christened. I was confirmed. I went to church a lot. I can still sing you lots of hymns and know all the words. I could probably recite most of the communion service by heart—I discovered that when my 80-year-old mother was very demented and unable to go to church on her own. So one day I took her to communion, and I found all the words just poured out of my mouth as though it hadn't been thirty years since I last went to a service.

I know quite a lot about Buddhism from a long training in Zen. I am not a Buddhist. I've not signed up, promised anything or taken the vows, but I have been practicing and training in Zen for more than thirty years now. I know just a little about Islam from having studied it for my own interest, just by reading, and rather less about Hinduism and much less about all kinds of other religions. So I am not speaking about them all.

Now, let's tackle these questions you ask about.

Cosmology. I'm not a physicist, I'm not a cosmologist, and I don't know where the scientific answers stand today, but I don't think we have a generally accepted answer to how and why the universe came into being. Even so, it doesn't help to invent a god who did the job. In the monotheistic religions this raises the classic infinite regress – Who created God? – as well as many other well-known problems. But there still remains the big question of why there is something rather than nothing. This is the sort of question one can think about religiously and spiritually and scientifically. I think the answer is not totally clear with cosmology. At least in my mind it isn't.

The other parts of your question I would like to give firmer answers to from my own perspective. As far as biology is concerned, science and religion are absolutely not compatible. In particular, this is true for Christianity, Islam and any religion which specifies a God who made us.

I have particular anger and despair about the Christian notion that God created us in his own image. You could take a rather subtle spiritual sort of notion of what that means and say "oh, well, it's for God's good qualities that we are made in his image. He wants us to love and forgive as he does; to aspire to spiritual perfection." But when you read the Bible or the Koran, you find that God encourages hostility to outgroups, glorifies war, approves of rape, and behaves in countless ways that we consider immoral today. He displays love and wrath, rejection and compassion, forgiveness and revenge in a soup of contradictions. All this may sum up our human failings rather well but it is far from what Christians mean when they say we are made 'in His image.'

A more down to earth meaning of 'made in His image,' entails what human beings look like, how our bodies are constructed and how we behave. The science of biology gives us answers. We know that we evolved along with all other living things on this planet. We may not know every detail of the processes involved but we know who our ancestors were, what evolutionary pressures were involved and where all the different genes come from. Religion cannot provide any such meaningful answers.

Then one can ask such difficult questions as: If God did create us, why on earth did he make such a horrible mess of it? Why are we so weak and so prone to becoming ill? Why are our backs not designed to carry so much weight? Why are our eyes not as sharp as an owl's? Why do our immune systems sometimes turn in on their own bodies? Why is childbirth so painful and dangerous when other animals find it easy and painless? Biology gives meaningful answers to all such questions. No religion offers any viable explanations at all. The Christian idea that we are made in God's image is simply ridiculous.

So my answer to your question about biology is no—and for a profound reason. In both Christianity and Islam God had a plan for us. He knew what sort of creatures we were to be and created us according to his plan. Evolution, by contrast, has no plan and no foresight. Had conditions been slightly different at any point, we would have turned out differently, or not appeared at all. This is its wonderful power— and where the deepest incompatibility between science and religion lies. When we look around the manmade world it seems just obvious that design requires a designer; that you have to know how to make some-thing in order to make it. It seems just crazy that beautiful and intelligent creations could appear through the aimless processes of copying with variation and selection. Yet this is what Darwin realized.

Design is indeed possible without a designer; designerless designs are all around us. We are designerless designs ourselves. There is no need for a designer God.

When it comes to ethics, I say no as well. Religions are often touted as providing a moral compass for their followers, but the morality displayed in the Old Testament or in the Koran is mind-numbingly gruesome. In the Old Testament, the Lord sends plagues of locusts, maggots and festering boils to his enemies. He condones selling girls into slavery and murders children. In one particularly nasty episode, "Moses' anger waxed hot" when the people disobeyed the Lord and made a golden calf. So he ordered his followers to kill their own brothers, friends and neighbours, "and there fell of the people that day about three thousand men" (Exodus 32:28). Similarly the Koran threatens horrific punishments to those who reject the faith, including death, crucifixion, and having their hands and feet cut off on alternate sides (5:33). It teaches that unbelief is worse than killing, and all unbelievers deserve to die.

The many stories of violence, theft and oppression in both 'Holy' books are understandable since the books were written by people, and the stories they tell expose common aspects of human nature. In that sense, there's a certain kind of compatibility between science and religion here. For example, science shows us that we humans are a social species having a common ancestor with chimpanzees, and other previous common ancestors with other social species. This helps us to understand a lot about our behaviour, including why we are sometimes altruistic and sometimes the reverse; why we are kind to our relatives and love our family members yet often shun outsiders; why racism is so easy to provoke given that there are genetic reasons for preferring people who are closely related to you or look like you. Psychology, neuroscience and memetics all provide more insight into how our cultures evolve and why we behave as we do. But it is science that explains all this, not religion.

The violence and horror of these 'Holy' books makes sense when you see Christianity and Islam as self-perpetuating memeplexes. Not only do they piggy-back on the human nature that evolution has given us but they cleverly exploit our minds to get themselves copied down countless generations. They can be seen as viruses of the mind that essentially consist of a set of 'copy-me' instructions backed up by all sorts of protective memes. So they urge people to copy all the senseless details of stories, rituals and doctrine, and then back that up with terrifying threats and false promises. And in case people don't believe the crazy threats and promises, these in turn are backed up with admonitions to have faith rather than doubt, not to ask questions about invisible heavens and hells and, above all, not to laugh. Laughing at the idea of a god who rewards his servants with a heaven full of sexually available virgins is surely going to weaken its power. So then laughing needs to be punished as well.

Using the tools of science we can understand how and why the world's nastiest religions evolved the way they did. And this understanding is incompatible with the teachings of those religions.

The next part of your question was about minds, brains and souls. I'm not a materialist, but I am a monist. I think when it comes to minds, brains and souls, we are never going to understand any of them by taking a dualist view. As Dan Dennett long ago said, "Dualism is forlorn," and almost every living philosopher would agree with him. If you separate the world into mind and body, or subjective things and objective things, or things you can measure and things that are only in your experience, then you have a problem. There's no easy way to understand how one relates to the other. If you take the Cartesian dualist view that mind and brain are completely distinct kinds of stuff, then you can't understand why putting a drug in your brain changes the way you see the world and you can't understand why your thoughts seem to relate to what your body does. This kind of dualism is hopeless.

I'll dismiss the idea of a soul completely because there is no separate entity that could count as a soul.

As for mind and brain, I think of mind loosely as what the brain does or an illusion constructed as part of the brain's job. What I think is particularly interesting—something you might add to minds, brains, and souls—is the nature of "self," because each of us, almost all of us, feels as though there is a "me" inside. The analogy of the driver and the carriage is a very common one in various religious and spiritual traditions. Indeed it feels as if "I" am in here, "I" am now waving my hand in front of my keyboard, and "I" caused that action. I can look out of the window to the trees being blown in the summer wind and think "I" am seeing the trees. "I" am in here and the trees are out there. All of this is illusory. The way it appears is not how it is. There's a brain and a body and a world that are all interacting with each other and somehow they give rise to this sense that I am separate from all of the rest of it. That is the illusion. There is no separate "me." There is no separate "self." The interesting question then becomes why does it seem that way? This is a question I am endlessly playing with, from both a scientific and a spiritual point of view.

Finally, we come on to free will and this is one of my great enthusiasms. I am not a philosopher. I am not qualified to talk in terms of all the complexities that thousands of years of discussion about free will have entailed. Indeed free will is said to be the most discussed philosophical issue of all time, and I can believe that. What I am interested in is how we can live our lives if we don't believe in free will.

I haven't believed in free will for many, many decades. I know that because I've recently reread my diaries from the 1970s. In my early twenties, I was arguing with people about free will and always saying that our will cannot be free. We can have a stronger or weaker will. Some people have a very strong will, but that comes down to their genes, memes and life experiences. Their strong will is still not free. Everything to do with will, everything to do with agency and action, is based on what went before in the brain, body and world.

If, for example, I decide now to thump the table, that is because of all the things that have gone before. Genes, memes, environment, the fact that you are listening, the fact that you asked me the question, the fact that it was tempting to join in your book, the fact that at that moment it seemed appropriate, for various reasons to do with my linguistic past, to think of something to do and the convenience of the table here under my hand. Everything is caused by what went before. That doesn't mean that it's predetermined, of course, but that it has causal antecedents. So where could the magical powers of free will possibly fit in? As with God, we have no need of this hypothesis.

Claiming that free will is illusory makes the most direct and wonderful clash with some religions and not with others and this I find really interesting. Take Christianity and Islam. In both of these religions, God gave us free will. Islam gets into complicated tangles over this issue with lots of arguments about predestination because, as so many Islamic scholars have noticed, there seem to be awful contradictions in the teachings. Generally speaking, though, there is a belief that human beings have free will even though nothing occurs except by Allah's decree.

In Christianity it is said that God gave us free will so that we could choose between good and evil. So here is this universe set up by God with good and evil in it, or perhaps God and the Devil, and we poor creatures are created in God's image. (That's odd because if we were in his image, wouldn't we always do the right thing?) It's set up so that we have to make this choice and we have free will to do so.

If you took away free will from that Christian Dogma, then you would find that there wasn't any sense in us choosing good or evil. Of course, this has implications for the way we understand our own society and, in my mind, people act well or badly because of the background that they've had. This doesn't mean that we should exonerate them from the consequences of their actions, but it does mean that we shouldn't judge them as inherently good or evil of their own free will, but as part of the universe of evolved beings behaving in the way they must in the circumstances they find themselves in.

Let me now turn to Buddhism because here you find a totally different view—particularly in Zen Buddhism. Now the trouble is that Buddhism, like all religions, covers all sorts of different ideas in many, many branches. I'm speaking here not from studying Buddhist doctrine but from my own thirty years training in Zen and long practice of meditating every day.

Deep down in the Buddhist teachings is one central doctrine called "codependent arising" or "dependent origination," and I think this is most marvelous. The Buddha, two and a half thousand years ago, said that everything that happens arises from what went before, which was quite contradictory to the folk culture in which he grew up. He rejected the popular ideas of spirits and invisible powers and said, no, everything happens because of what happened before.

There is no specific doctrine of free will within Buddhism. Yet that central idea of codependent arising seems to me to do away with free will. There are also many practices which involve 'letting go' or 'letting be' and not trying to control the mind. For example, central to many meditation practices is the idea of letting go of clinging to all sorts of things, but in particular clinging to self—then you see that it is the illusory self, or constructed self, who would be the one to exert the free will. If you let go of that, then free will disappears too.

It's often said that the ultimate in Zen, the ultimate conclusion of long years of meditation, is non-meditation. I must say, when I first heard that, probably twenty years ago, I was completely horrified— "You mean I'm going to spend years and years meditating? It's so painful and I hate it, and I sit on my cushion and my legs hurt. I hate the thoughts I have. It's all so absolutely ghastly and what I'm heading for is non-meditation? Ugh!"—it seemed such a counter-intuitive and horrific idea. And yet now I'm very comfortable with that idea because when the self who would be doing things slips away, the question arises who is meditating? This question can be used as an especially difficult koan; a question to meditate with, but after a while it simply loses its power. Any thought that "I" am meditating seems no longer to have the grip that it used to have. Meditation happens. This is all quite different from the teachings of the monotheistic religions.

My absolute favorite all time quote from the Buddha is, "actions exist and also their consequences, but the person that acts does not." I think we can see that in that religion at least, the ideas of science and religion can be compatible.

3. Some theorists maintain that science and religion occupy non-overlapping magisteria—i.e., that science and religion each have a legitimate magisterium, or domain of teaching authority, and these two domains do not overlap. Do you agree?

No. I completely disagree, particularly with Steven J. Gould's explanation of the separate magisteria, for the reasons that I've given you in the answer to question two. Various religions and science ask questions about the same things. They ask questions such as: Who am I? Why am I here? What is the universe made of? Do I have free will? All these are questions that both science and religion try to answer. Science gives better and better answers as it continues delving into the universe. Religion just gets itself in ever deeper tangles. I think it's a cop-out to say they have separate magisteria, and thereby somehow give religion a space to occupy that it doesn't deserve.

4. What do you consider to be your own most important contribution(s) to theorizing about science and religion?

Memetics. How shall I begin this? I love the idea that Richard Dawkins started in his book *The Selfish Gene* that we can look on habits and skills and stories and songs, all of the things that we pass from person to person and culture to culture, as *memes*—a kind of cultural equivalent of genes.

A lot of people hate memetics. There are very, very few scientists who agree with me about its potential power. But I think there are more and more scientists who think that culture evolves as opposed to being, like Steven J. Gould who said, "I wish people would stop talking about cultural evolution." Culture does evolve, and I like to think of that process in terms of the competition between memes to use us human meme machines to get themselves copied.

Religions are absolutely the epitome of a selfish memeplex. *Memeplex* is an abbreviation of 'co-adapted meme complex.' In other words, a whole lot of memes that hang around together because they are more likely to get passed on from person to person, (or person to book, or book to computer) as a group than they would on their own. You could say that they are protected by the informational equivalent of a membrane and create something like an organism.

So why do I think of religions like this? Religions have much in common with email viruses, chain letters and so on because their essential structure is a copy-me-instruction backed up by threats and promises.

In almost any religion, you will find that you are told by God or by the priests or by whoever it may be, that you are to spread the religion. Sometimes it's in a cheerful version, "Spread the good news of Jesus, we're all going to be saved!" or it can be in a horrible version, "If you don't spread this idea, if you don't bring your children up in this religion, you'll be hacked to death, ostracized, murdered, or tortured." If the threats are for punishment in this life, as they are in Islam, it is understandable that people conform, and so the religion continues on its way. If they are for the life hereafter then the memeplex needs further tricks to make that mythical place believable, such as promoting faith over doubt, banning any attempt at testing, and treating laughter as blasphemy.

There are lots and lots of other tricks that religions use to get themselves copied. Memes are just information: Information that can manifest itself in the form of writing in a book, digits in a computer, words spoken by a human, behaviors that can be copied like eating with a knife and fork or driving on the left or the right of the road. They are information that replicates and can keep changing form as it competes with other information to fill up the niches in culture.

There are obviously many niches for religions to compete for in the human mind. We long for answers to difficult questions including the ones we've been discussing here. We long to know: What should I do now? What is right and what is wrong? Why is the world so unfair? Why is there so much wickedness and suffering? Religions appear to give answers, so that's one niche that they're filling. They help some people to face death and sickness by offering life after death. They promise to even up the unfairness of life by rewarding the good and punishing evil in the invisible hereafter.

Many religions do this but how and why does one religion out-compete another? We might imagine the 2,000-year history of human religious development and consider all the millions of little cults there must have been. Jesus started out with twelve disciples. There must have been many other charismatic leaders gathering little groups of people around them. Most of these proto-cults would have died out with the death of their leader, and this probably happened again and again and again over thousands of years. Some of them led to Christianity and all its many, many branches, or to Islam, or to different branches of Hinduism and so on. The development and division of the world's religions have been mapped out in great detail and really do form evolutionary trees. So why did these few religions spread across the world and the rest die out?

I would say not because they are true or useful or give valid answers to the questions that science is now answering properly. I would say it's primarily because of the tricks they use. These are not just threats and promises. Religions also use what I call the beauty trick. If you go into a beautiful church...and goodness me, there are the most wonderful churches around here. I live in a small South Devon village with an amazing thirteenth-century church. It's absolutely beautiful and there are many others like it around here. Every time I go inside I almost draw a breath at the calmness, the quiet, the ancient stone, the stained glass windows, the beautiful altar, the carvings. You can just stand there and be transfixed.

That is true beauty. That is true wonder. But of course while you're there, you cannot avoid the images of Jesus Christ and the Virgin Mary, the illustrations of bible stories, and the reminders of the threats and promises that Christianity offers. If you're there for a service, you cannot escape the teachings. I was at a funeral last week and was horrified when the vicar started shouting loudly about how we Christians know that death is not the end and our beloved here (whose coffin was in front of us), will be going on to the many rooms in God's mansion. All this gibberish is more likely to be believed because of the beauty, the serenity, and the wonder of the surroundings. Then there's all the wonderful, glorious music and the beautiful words of the Kings James Bible—which I do think has beautiful words in it even if I dislike the content so much. And there are the catchy tunes of hymns that have been selected over centuries to take their place in religious services— tunes that many of us learn as children and can never forget.

Then there's the altruism trick. This is really nasty because almost all religions, and certainly Christianity, tell their believers that if they follow the teachings they are good. They tell people—you are a good person because you're a Christian and you do as God wants (of your own free will of course!). The implication is that other people are not so good and you are morally superior to the unbelievers. Many Christians are indeed kind and generous people. You can ask them why and they often say that the reason they do helpful things, or give money to charity or volunteer in their community is because of their faith. It is because they follow Jesus that they do this, or for Muslims it is because they worship Allah that they do these good things.

But I'm happy to say there are recent studies done on the actual altruistic behavior of believers and atheists. When comparing levels of prosocial behaviour there is no consistent difference between atheists and religious people. It is just that the atheists cite their own conscience or personal values rather than their faith as the reason for doing good.

The altruism trick is extremely powerful and is found in almost every religion, promoting empathy, compassion, kindness, and above all, love. Of course the love and kindness can be genuine but I call this a trick because religions use all their talk of love and compassion to further their own agenda. And why does this work? Because we want to be good. And why do we want to be good? Because of our evolved nature. If you are seen to be good, whether you are a chimpanzee or a human being or one of many other social species, you gain an advantage. You may gain status, get a better mate, and get better food, if other people like you. If others think you're trustworthy and generous they will want you in their group and will be willing to return your favours in the game of reciprocal altruism.

There are many other tricks by which religions piggy back on human nature and lure people into adopting their beliefs, customs and regulations. And they add to them all by demanding money to support the spread of those religious ideas. This is made to seem like yet another altruistic action, when all it does is trick yet more people into believing nonsense and spending their precious time and money promoting these religious memes. The gist of all this is to say that if I've made any contribution towards the relationship between science and religion it is to take the idea of memes seriously and show just how and why religions can control people's lives despite being riddled with nonsense.

I would like to add here one other idea to do with memes and religion, which specifically concerns Buddhism. What I find really fascinating is how strongly the popular idea of Buddhism revolves around a really crude version of personal reincarnation. This is the idea that each person is reincarnated into another body or a whole series of bodies, and might either descend in the next life to become a frog or ascend through higher levels and evolve spiritually as the highest kind of being—a human who can become a Bodhisattva. This is particularly prevalent in Tibetan Buddhism, possibly because reincarnation was already part of the existing folklore of Tibet when Buddhism first arrived there.

Indeed, all over the world, people have invented the idea of reincarnation again and again. It's just one of those natural kinds of ideas that our brains are suited to. It comes easily from the false idea of the self who lives inside the body and from asking—When I die, what happens to me? My body rots, or is burned or whatever, but what happens to "me"? Some religions answer by saying that "I" go on to live forever in heaven or hell; others that "I" am reincarnated as something else. This is tempting to believe for surely "I" cannot just not exist after death.

The weird thing is that if you read what the Buddha said, you find that the central doctrines of Buddhism are all about no self, or *anatta*. This doesn't literally mean there is no self but that the self is not what it seems to be. In other words "I" am not a persisting entity. The self is not something that has experiences or exerts power or controls its body. It's an ephemeral construct that arises and falls away and arises and falls away. There may still be some kind of continuity because of memory but this comes down to the continuity of the body—not of some separately existing inner being. Death in this context becomes no different from life because in every moment of life selves are arising and falling away. So death will be just another falling away.

This is indeed a tough doctrine to swallow. So let's compare these two ideas. First there is the idea of personal reincarnation, that "I" will not disappear when my body dies but will go on to live another life. Everything I do now will have meaning and purpose because of its consequences in that future life. The alternative view of *anatta* is that nothing ultimately matters, for everything is ephemeral, including my precious self. And look—that self is already gone. Here's another me! And what about the one who started trying to answer these questions at the beginning? She's long gone. All selves are ephemeral constructs and really of no consequence.

Which idea is going to be more popular? Which meme is going to get itself copied into more brains? The answer is surely obvious. The first is a much stickier meme than the second. It usually takes years and years and years of meditation and mindfulness for people to become comfortable with the idea of letting go of self—of accepting that self is an ephemeral construct in an ever-changing universe devoid of ultimate purpose. So when these two different memeplexes compete in our modern world, the first (false) one thrives. Like science, the Buddha's ideas challenge everything we like to believe. For this reason they do not make popular memes.

This leads me to one final part to my answer, which concerns spirituality. For all I've said against religions and the horrors they commit, I think most of them have, deep down in their core, deep spiritual truths. I have difficulty with that word "spiritual," because it implies there must be such a thing as a spirit. I don't mean that at all and yet I don't have a better word for it. I mean something that leads some of us to pursue meditation, to train our minds to become gentle and peaceful, to have mystical experiences in which the self dissolves into the universe, to allow experiences of non-duality.

Right in the heart of Christian mysticism, and in Sufism, and in the Advaita tradition of Hinduism, you find the same ideas of letting go of self, of being one with the universe, of surrender to what is. Yet the popular versions of all these religions are quite different. I think we can therefore see the process I described in terms of Buddhism and reincarnation as quite general. Difficult spiritual insights get washed away by popular ideas of self and gods backed up with clever tricks.

If it is true that there is no persisting self, no free will, no stream of conscious experiences, and the universe is just as it is, that's tough for any of us to accept. So the popular memes of God, the afterlife, souls, and spirits that form such an important part of religion, will nearly always win out. Nearly always! Not quite. I hold out some hope that maybe they won't always win.

5. What are the most important open questions, problems, or challenges confronting the relationship between science and religion, and what are the prospects for progress?

I was at a conference last week and there was a cloth bag for sale which said "Religion! Together we can find the cure." I thought that was absolutely wonderful. That to me expresses the problem or the challenge confronting us now.

To put it in memetic terms, we need to strengthen memetic immunity. We need to find or create successful memes that will help people to look critically at things they have previously just accepted. Almost every religious person alive today holds to the religion of their parents. This indoctrination when they're very young makes the religion hard to throw off but it must be possible to find ways to help people throw off their religious beliefs if they find them wanting.

This can be hard when you consider some of the consequences of rejecting one's religion. In Islam apostates may be rejected by their families, thrown out of their community, and threatened with injury, mutilation, or death if they give up their faith. Surely, there are counter memes. Surely there are disinfectant memes that might spread and help people escape the tricks religion has played on them. To some extent this is already happening. I think the opening up of the World Wide Web and the international media means that people are not kept totally ignorant of other religions. And that really helps.

I reflect on the situation in my children's school and how different things are in the English education system from the American education system. It's really strange and perverse. In the United States, you have the idea of separation of church and state, which means that you're not allowed to teach religion in schools. Here in England, we have a state religion. The Queen is the Head of the Church of England. That is our nation's religion and we also have compulsory religious education in schools.

What has been the consequence of this? In the United States, the majority of people are religious. Politicians mention God. They say "God bless you" in their speeches, and "God be with us," and awful things like that. Children are generally brought up — because they have no religious education in school — knowing only about the religion of their parents. They're taken to a church or mosque or synagogue and taught that their own religion is true from a young age. This makes it very hard to throw them off.

By contrast here in England, we have compulsory religious education and, because we have a lot of different people from different religions living here, that means that religious education has become comparative religion. My kids are well grown up now, but I remember them coming home from school with some friends and they were larking around and saying, "Guess what we learned today? The Sikhs wear turbans on their head and they believe you have to have a sword. Do you know Muslims believe that . . ." They were all kind of throwing these ideas around and they must have been seven or eight years old. It doesn't take a very brainy kid who learns about a whole lot of religions to come to the conclusion that they can't all be true. And that is the starting point of being critical and escaping from the feeling that whatever you were taught as a child must be the truth.

We get on fine with our state religion in England by and large. The majority of people now in surveys do not describe themselves as practicing any religion. In the last census, we had thousands of people describing their religion as Jedi, or the flying spaghetti monster. More important, about a quarter of the population of England and Wales reported "no religion." This is not a country where belief in God thrives, or where being an atheist is considered remotely odd or unusual. Indeed I don't know or meet many people who are traditionally religious. I once asked my kids whether they knew any religious people in their school. My son pulled a face and said, "Yes, there are two girls in our class who believe in God!"

There is hope, but that's the challenge. The challenge is to help people to escape from the oppression of the religion they grew up with. And we need to do that in such a way that doesn't oppress those who still want to believe, that doesn't throw out natural human curiosity or human awe, and doesn't throw out what I have to call, because of lack of a better word, our spirituality, our spiritual nature. I hope that we can retain the capacity to follow a spiritual path, to practice meditation, contemplation or mindfulness, to work towards a different relationship between self and others. I hope we can strive to become more compassionate, more understanding and freer from the oppressions that the memes put on us all the time.

These really are challenges. And the prospects for progress? I don't think I'm going to answer that one, but I do remain hopeful.

3

Sean Carroll

Sean Carroll is an American theoretical physicist and writer. His research focuses on theoretical physics and cosmology, especially the origin and constituents of the universe. He has contributed to models of interactions between dark matter, dark energy, and ordinary matter; alternative theories of gravity; and violations of fundamental symmetries. Carroll is the author of *Spacetime and Geometry: An Introduction to General Relativity* (2003), *From Eternity to Here: The Quest for the Ultimate Theory of Time* (2010), and *The Particle at the End of the Universe* (2012). He has been awarded fellowships by the Sloan Foundation, Packard Foundation, and the American Physical Society, as well as the Royal Society Winton Prize. Carroll has appeared on TV shows such as *The Colbert Report* and *Through the Wormhole with Morgan Freeman*, and frequently serves as a science consultant for film and television.

---

1. What initially drew you to theorizing about science and religion?

Anyone interested in either science or religion should care about the relationship between the two. Both make strong claims about the way the world works—claims that many people accept, on both sides. But claims that (at least for certain definitions of "religion") seem to be at odds. I don't actually spend much time theorizing about science and religion. I think about how the world works, and try to draw reliable conclusions from the evidence we have. My own conclusion is that the fundamental ontology of the world is naturalistic: there is only one world, the natural one, whose behavior is studied by science. This has *enormous consequences* for everyday human life, and in particular for questions that are often discussed in the context of religion.

Is there life after death? No, because we know enough about physics and biology to understand that there's no way the information characterizing a human soul can exist outside the body. Is there a purpose to the universe? There certainly isn't any purpose evident in the fundamental laws of physics, which are perfectly mindless and dysteleological. Is there a natural way of being that can be used as a normative guide to proper human behavior? No, there are only atoms and particles doing what they do.

I think it should be the responsibility of people who professionally think about the ultimate nature of reality to share the implications of their work with the broader community of people who are interested in questions like these. Insights into foundational science and philosophy don't exist in a vacuum, and shouldn't be relegated to the ivory tower. An enormously successful worldview has been put together in the years since Galileo, but it hasn't yet been accepted by most people, even well-educated ones. I'd like to change that, since we need to start from agreed-upon ground in order to make progress on difficult questions where agreement is hard to come by. That's why I talk about science and religion.

2. Do you think science and religion are compatible when it comes to understanding cosmology (the origin of the universe), biology (the origin of life and of the human species), ethics, and/or the human mind (minds, brains, souls, and free will)?

The question of the compatibility of science and religion hinges crucially on what one means by "religion," which is hardly uncontested territory.

One move is to simply define religion as "questions of meaning and human value" (regardless of one's answers), in which case science and religion are manifestly compatible. But I don't think that's actually what most people mean by the term. And it's horribly unfair to real religious belief and practice, which generally involves much more than the mere existence of a set of values.

Another possibility is to think of religion as a route to knowledge that is distinct from science—one based on revelation and contemplation rather than on empirical investigation. That conception seems as straightforwardly incompatible with science as the previous one seems compatible. History teaches us that the only way to get reliable information about how the real world works (as opposed to purely logical truths) is some version of the hypothetico-deductive method: propose ideas, test them against the data.

But the interesting (to me) version of "are science and religion compatible?" is the one that compares their substantive claims about the world. There are some traditions that are arguably "religious" but also naturalistic—some versions of Buddhism or Unitarianism. But the large majority of religious believers are not naturalists; they think there is something real other than the natural world, whether it be God or spirits or merely purposes. When it comes to those beliefs, I think they are straightforwardly incompatible with the findings of modern science. Not with the *methods* of science—I see no obstacle in principle to scientists looking at the world, applying their hypothetico-deductive techniques, and coming back to report "God exists" or whatever. But that's not what has actually happened.

When it comes to cosmology, biology, or the human mind, there is essentially no reason to give weight to religious traditions, any more than we would give weight to the scientific ideas of Aristotle or Avicenna. They may have been right about some things, and are certainly worth studying for historical or literary value and occasional philosophical insights. But when modern science disagrees with them about how the world works, go with the science every time.

Ethics is different—that's a matter of value judgment, not merely about describing the world. Science by itself is insufficient to address issues like ethics, morality, meaning, or purpose, because science is resolutely descriptive rather than judgmental. It's even conceivable that we might find useful suggestions in the texts and beliefs of religious traditions. But there's no sense in which those suggestions should be given extra weight simply because they arose from religion. Ethics comes from human beings working out how best to reconcile their individual and collective desires for fairness and communal activity, not from instructions handed down by God.

3. Some theorists maintain that science and religion occupy non-overlapping magisteria—i.e., that science and religion each have a legitimate magisterium, or domain of teaching authority, and these two domains do not overlap. Do you agree?

Not in the slightest, as should be obvious from the above. The non-overlapping magisteria (NOMA) idea was popularized by Stephen Jay Gould in his book *Rocks of Ages*. It's hard to understand at first blush: science leads us to the conclusion that the world operates according to naturalism, while religion (typically) posits the opposite, so that sounds like a pretty obvious overlap.

The answer to this puzzlement is that Gould was using a bizarre definition of "religion." He writes, "This magisterium of ethical discussion and search for meaning includes several disciplines traditionally grouped under the humanities—much of philosophy, and part of literature and history, for example. But human societies have usually centered the discourse of this magisterium upon an institution called 'religion.'"

In other words, he wasn't talking about religion at all, in the way it is usually understood in the Western world. He was talking about "ethical discussion and [the] search for meaning." It's a version of "religion" that doesn't include the existence of God, the possibility of an afterlife, or other supernatural claims. It's a common move among people who aren't traditional believers themselves, but want to be sympathetic to some benign formulation of religion: they water it down by ignoring some of its most defining features. It strikes me as very strange indeed, especially on the part of a scientist, to treat truth claims about how the universe works at a fundamental level as somehow unimportant or irrelevant.

Whatever human societies may usually have done, there is no reason whatsoever to couch the discourse of ethics and meaning in religious terms. Given a naturalistic view, the grounding of various claims about ethics and meaning will generally be very different than they are within religious conceptions. Believers can sensibly hold that a certain kind of behavior is unethical because God said so, or because it goes against the ultimate purpose of human existence. Naturalists aren't able to make such claims, but that doesn't mean they can't talk about ethics or meaning; it's just that their ultimate justifications must be very different.

Even if we define the two magisteria as "science" and "values," it's still an exaggeration to pretend that they are *strictly* non-overlapping. Science is not sufficient to determine ethics, since all it does is describe how the world works. But it is certainly necessary. If we don't have an accurate picture of how the world works, it would be very hard indeed to settle on a sensible theory of ethics. One's attitude toward the legality of abortion, just to choose one topical question, might very well be greatly influenced by whether or not there exists a life force that enters the fertilized ovum at the moment of conception. Similarly, questions about animal rights, genetically modified foods, or psychotherapeutic techniques are obviously informed by science. It's important to appreciate the fundamental differences between science and values, but a mistake to think that they are completely separate from one another.

4. What do you consider to be your own most important contribution(s) to theorizing about science and religion?

I'm not sure I've contributed anything extremely important to the area, to be honest. Most of the arguments I have put forward amount to reiterating points that have been discussed for decades, if not centuries. As a working theoretical cosmologist, I'm able to be a bit more specific than the average atheist about why arguments from fine-tuning or the need to understand the origin of the universe don't make a persuasive case for God. But I'm hardly the only one.

One point that I do like to stress, which isn't made elsewhere, is how much we already know about the fundamental workings of reality, and how much our knowledge constrains novel possibilities. This is a delicate subject, because (as I've learned) people are very inclined to misunderstand you in one direction or another. There's obviously a lot we don't understand about the universe, from dark matter to the Big Bang. And (even more obviously) there is an enormously larger amount we don't understand about the actual workings of the emergent macroscopic world, from turbulence to neuroscience to stock markets. Science is nowhere near finished.

But there are, nevertheless, things that science does pretty much understand—theoretical models that we believe are correct within a certain well-defined domain of applicability, and which will continue to be correct no matter what else we learn about the universe. (Newton's view of gravity was superseded by Einstein's general relativity, but it continues to be an accurate description of what happens in a certain regime; the phlogiston theory of combustion, by contrast, was simply wrong.)

And among the things we do understand is the behavior of the microscopic constituents of everyday matter—the particles and forces that make up you and me, the Earth and Sun and stars, and everything we've ever directly detected in a laboratory. The combination of the Standard Model of Particle Physics with general relativity (with very minor adjustments to allow neutrinos to have mass) is perfectly consistent with every experiment we've ever done here on Earth. More importantly, while this theory is certainly not the final answer, the basic structure of quantum field theory ensures us that any deviations from this model will be completely irrelevant to our everyday lives. There can be new particles, but they must be too difficult to produce or too short-lived to matter, or we would have detected them already. And there can be new forces, but they must be too short-ranged or weakly-interacting to have any appreciable affect on the atoms of which we are made, or again we would have found them already. The physics underlying the everyday world is completely understood.

That's bad news for anyone who wants to believe in extra-physical entities, up to and including an immaterial soul that survives the body after death. It's obviously not a proof that such entities don't exist, since such a proof is impossible (you can always imagine anything you like that conspires to hide itself from all experimental inquiry). But it highlights exactly *how* dramatic a departure from known reality is required to hold on to such ideas in this day and age. We no longer live in the world of five hundred years ago, when the motion of the planets was mysterious and it was natural to think that living beings were somehow fundamentally different than non-living matter. We are made of atoms, and we have a successful theory of how those atoms behave from which there is precisely zero evidence of any kind of deviation.

Obviously (it's important to keep saying "obviously," given people's desire to misconstrue this point) we don't understand the precise way those atoms come together to make a human being. But no matter how that happens, there's no room in our successful theory of physics for any fundamentally new influences acting on the atoms themselves. The relationship between atoms and macroscopic reality is a question of emergence and collective action, not one of new basic entities. Believing otherwise isn't simply a matter of filling in gaps in our ignorance;

it requires that well-established physical theories are straightforwardly false, against all evidence. To me, this insight is as important as Darwinian natural selection in establishing the sufficiency of a naturalistic worldview.

5. What are the most important open questions, problems, or challenges confronting the relationship between science and religion, and what are the prospects for progress?

As one will have gathered from the above, I'm not primarily interested in "progress on the relationship between science and religion." I am, however, extremely interested in understanding how naturalism confronts some of the big questions that were formerly tied to religion. Given that we human beings are just complicated collections of atoms, we need to develop a more sophisticated understanding of concepts like "meaning" and "responsibility" and "morality." These aren't easy questions, and I don't think we're anywhere near having fully convincing answers.

We have plenty of *unconvincing* answers. Even the most committed naturalists seem willing to accept that we need to find absolutely secure and objective groundings for questions of judgment and morality, just that such grounding is supposed to reside in the natural world rather than in God or spirituality. But the natural world is completely nonjudgmental; it doesn't demand any particular purpose for human life, it doesn't classify some behaviors as moral and others as immoral, and it doesn't provide objective standards for beauty or aesthetic quality. It simply *is*.

That doesn't mean that there is no such thing as morality, or beauty, and so on. It just means that they're not objectively grounded in the natural world. Baseball isn't objectively grounded in the natural world, either; it was invented by human beings. That doesn't mean baseball doesn't exist, or isn't real.

The lesson of naturalism is that matters of judgment are ultimately human creations. That leaves a lot of work for us to do—what rules of morality and meaning are the best ones to create? Different people will answer the question differently, and that's okay. (It needs to be okay, since that's how the world is.) But there is still a good deal of hard work to be done to invent rules that satisfy our individual desires, as well as fitting in with the desires of others. Naturalism is an ambitious project, one that is still in its very early stages. I'm optimistic that human beings are on the road to getting it right.

4

William Lane Craig

William Lane Craig is Research Professor of Philosophy at Talbot School of Theology and a world-renowned philosophical theologian. He specializes in philosophy of religion, metaphysics, and philosophy of time. Craig has authored over a hundred and fifty professional articles and authored or edited over forty books, including *The Kalam Cosmological Argument* (1979), *The Cosmological Argument from Plato to Leibniz* (1980), *Divine Foreknowledge and Human Freedom* (1990), *Theism, Atheism and Big Bang Cosmology* (1993), and *God, Time, and Eternity* (2001).

---

1. What initially drew you to theorizing abtout science and religion?

As a boy, I was fascinated by science, particularly astronomy and paleontology, and participated in science fairs and other competitions. After becoming a Christian in my late teens, I naturally wondered how my scientific knowledge fit in with my new set of beliefs. One of the principal emphases of Wheaton College, which I attended, was the articulation of a Christian *Weltanschauung*, in which the input of theology was integrated with the results of other fields of knowledge, including the physical sciences. So it was at Wheaton that my interest in the integration of science and theology was quickened. This interest was augmented during my doctoral studies in philosophy, as I explored the implications of modern cosmology for the *kalam* cosmological argument's second premise that *the universe began to exist*.

2. Do you think science and religion are compatible when it comes to understanding cosmology (the origin of the universe), biology (the origin of life and of the human species), ethics, and/or the human mind (minds, brains, souls, and free will)?

Contemporary *cosmology* is one of the areas of science which is most compatible with traditional theism. Prior to the twentieth century the prevailing view *for millennia*, from ancient Greek philosophy through modern materialism, was that the universe is past-eternal. The Judeo-Christian view that the world was created a finite time ago stood against the received view of physics through the 19th century. The discovery of the expansion of the universe turned the received worldview on its head. The prediction of an absolute beginning by the standard big bang model of Friedmann and LeMaitre has stood now for over 90 years through the most tumultuous period in the history of physics. One challenge after another to the prediction of the standard model, from the steady-state model to oscillating models to vacuum fluctuation models to quantum gravity models, has succumbed to the evidence for an absolute beginning. Meanwhile, singularity theorems such as the Hawking-Penrose theorems, the Borde-Guth-Vilenkin theorem, and the Wall theorem continue to point to the finitude of the past and the beginning of the universe. Vilenkin has said, "All the evidence we have says that the universe had a beginning." This is clearly compatible with the Judeo-Christian doctrine that "In the beginning God created the heavens and the earth" (Gen. 1.1).

Moreover, during the last 50 years or so, scientists have been stunned by the discovery that the existence of intelligent, interactive life depends upon a complex and delicate balance of initial conditions given in the Big Bang itself. Scientists once believed that whatever the initial conditions of the universe, eventually intelligent life might evolve. But we now know that our existence is balanced on a knife's edge. The existence of intelligent life depends upon a conspiracy of initial conditions which must be fine-tuned to a degree that is literally incomprehensible and incalculable.

This fine-tuning is of two sorts. First, when the laws of nature are expressed as mathematical equations, you find appearing in them certain constants, like the gravitational constant. These constants are *not* determined by the laws of nature. The laws of nature are consistent with a wide range of values for these constants. Second, in addition to these constants there are certain arbitrary quantities which are just put in as initial conditions on which the laws of nature operate, for example, the amount of entropy in the universe. Now all of these constants and quantities fall into an extraordinarily narrow range of life-permitting values. Were these constants or quantities to be altered by less than a hair's breadth, the life-permitting balance would be destroyed and life would not exist.

There are three alternatives proposed in the literature for explaining the presence of this remarkable fine-tuning of the universe: physical necessity, chance, or design. The question is: Which of these alternatives is the most plausible?

The first alternative seems extraordinarily implausible because, as I said, the constants and quantities are independent of the laws of nature. For example, the most promising candidate for a Theory of Everything (T.O.E.) to date, super-string theory or M-Theory, allows a "cosmic landscape" of around 10^{500} different universes governed by the present laws of nature but with varying values of these constants and quantities, so that it does nothing to render the observed values physically necessary.

So what about the second alternative, that the fine-tuning of the universe is due to chance? The problem with this alternative is that the odds against the universe's being life-permitting are so incomprehensibly great that they cannot be reasonably faced. Even though there may be a huge number of life-permitting universes lying within the cosmic landscape, nevertheless the number of life-permitting worlds will be unfathomably tiny compared to the entire landscape, so that a dart thrown randomly at the cosmic landscape has no meaningful chance of striking a life-permitting world.

In order to rescue the alternative of chance, its proponents have therefore been forced to adopt the hypothesis that there exists an infinite number of randomly ordered universes composing a sort of World Ensemble or multiverse of which our universe is but a part. Somewhere in this infinite World Ensemble finely-tuned universes will appear by chance alone, and we happen to be one such world.

There are, however, at least two major weaknesses of the World Ensemble hypothesis: First, there's no independent evidence that such a World Ensemble exists. No one knows if there are any other universes at all. Moreover, recall that Borde, Guth, and Vilenkin proved that any universe in a state of continuous cosmic expansion cannot be infinite in the past. Their theorem applies to the multiverse, too. Therefore, since its past is finite, only a finite number of other universes may have been generated by now, so that there's no guarantee that a finely-tuned world will have appeared anywhere in the ensemble.

Second, and more fundamentally, if our universe is just a random member of an infinite World Ensemble, then it is overwhelmingly more probable that we should be observing a much different universe than what we in fact observe. Roger Penrose has calculated that it is inconceivably more probable that our solar system should suddenly form by the random collision of particles than that a finely-tuned universe should exist. (Penrose calls it "utter chicken feed" by comparison.) So if our universe were just a random member of a World Ensemble, it is inconceivably more probable that we should be observing a universe no larger than our solar system. For there are far more observable universes in the World Ensemble in which our solar system comes to be instantaneously through the accidental collision of particles than universes which are finely-tuned for intelligent life. Indeed, the most probable observable universe is one in which a single brain fluctuates into existence out of the quantum vacuum and observes its otherwise empty world. Observable universes like those are just much more plenteous in the World Ensemble than worlds like ours and, therefore, ought to be observed by us. Since we do not have such observations, that fact strongly disconfirms the multiverse hypothesis. On naturalism, at least, it is therefore highly probable that there is no World Ensemble.

It seems then that the fine-tuning is not plausibly explained by physical necessity or chance. Therefore, we ought to prefer the hypothesis of design, unless the design hypothesis can be shown to be just as implausible as its rivals. Detractors of design sometimes object that on this hypothesis the Cosmic Designer himself remains unexplained. It's said that an intelligent Mind also exhibits complex order, so that if the universe needs an explanation, so does its Designer. If the Designer does not need an explanation, why think that the universe does?

This objection is based on a misconception of the nature of explanation. It is widely recognized that in order for an explanation to be the best, you don't need to have an explanation of the explanation. For example, if astronauts were to find a pile of machinery on the back side of the moon, they would recognize that it was the product of intelligent design even if they had no idea whatsoever who made it or left it there.

In order to recognize an explanation as the best, you don't need to have an explanation of the explanation. In fact, when you think about it, such a requirement would lead to an infinite regress of explanations, so that nothing could ever be explained and science would be destroyed! So in the case at hand, in order to recognize that intelligent design is the best explanation of the fine-tuning of the universe, one needn't be able to explain the Designer. Whether the Designer has an explanation can simply be left an open question for future inquiry.

Moreover, the complexity of a Mind is not really analogous to the complexity of the universe. A mind's *ideas* may be complex, but a mind itself is a remarkably simple thing, being an immaterial entity not composed of pieces, or separable parts. Moreover, properties like intelligence, consciousness, and volition are not contingent properties which a mind might lack, but are essential to its nature. Thus, postulating an uncreated Mind behind the cosmos is not at all like postulating an undesigned cosmos with all its contingent and variegated quantities and constants.

Thus, contemporary cosmology is remarkably compatible with the belief in a Creator and Designer of the universe.

Contemporary *biology*, and in particular evolutionary theory, is pretty clearly compatible with generic theism. Some people have disagreed with this judgment because according to the neo-Darwinian theory of evolution the mutations which serve to drive the evolutionary process forward are *random*, from which they infer that they cannot be designed or occur for a purpose. But this inference involves a fundamental and very important misunderstanding about what biologists mean by the word "random." When biologists say that the mutations responsible for evolutionary change occur randomly, they do *not* mean by chance or purposelessly. If they did, then evolutionary theory would be enormously presumptuous, since science is just not in a position to say with any justification that there is no divinely intended direction or goal of the evolutionary process. How could anyone say on the basis of scientific evidence that the whole scheme was not set up by a provident God to arrive at *homo sapiens* on planet Earth? How could a scientist know that God did not supernaturally intervene to cause the crucial mutations that led to important evolutionary transitions, for example, the reptile to bird transition? Indeed, given divine middle knowledge, not even such supernatural interventions are necessary, for God could have known that were certain initial conditions in place, then, given the laws of nature, certain life forms would evolve through random mutation and natural selection, and so He put such laws and initial conditions in place. Obviously, science is in no position whatsoever to say justifiably that the evolutionary process was not under the providence of a God endowed with middle knowledge who determined to create biological complexity by such means. So if the evolutionary biologist were using the word "random" to mean "undesigned" or "purposeless," evolutionary theory would be philosophy, not science.

But the evolutionary biologist is *not* using the word "random" in that sense. According to the eminent evolutionary biologist Francisco Ayala, when evolutionary biologists say that the mutations that lead to evolutionary development are random, they do not mean "occurring by chance." Rather they mean "irrespective of their usefulness to the organism." Now this is hugely significant! The biologist is not, despite the impression given by partisans on both sides of the divide, making the presumptuous philosophical claim that biological mutations occur by chance and, hence, that the evolutionary process is undirected or purposeless. Rather he means that mutations do not occur for the benefit of the host organism.

If we take "random" to mean "irrespective of usefulness to the organism," then randomness is not incompatible with direction or purpose. For example, suppose that God in His providence causes a mutation to occur in an organism, not for the benefit of the organism, but for some other reason (say, because it will produce easy prey for other organisms that He wants to flourish). In such a case, the mutation is both purposeful and random. Thus, there is no incompatibility between generic theism and evolutionary biology.

How, then, are theological and scientific perspectives on the origin of life and the evolution of biological complexity best to be integrated? Here the theist, in contrast to the atheist, is free to follow the evidence where it leads. It seems to me an open question as to the extent of God's miraculous involvement in the origin of life and evolution of biological complexity. Most of us were probably taught in high school or elementary school that life originated in the so-called primordial soup by chance chemical reactions, perhaps fueled by lightning strikes. All of these old, chemical origin of life scenarios have broken down and are now rejected by the scientific community. Today there is a plethora of competing, speculative theories with no consensus on the horizon. The origin of life on earth remains scientifically inexplicable as things now stand.

Moreover, the odds of getting even a single functioning protein molecule by chance is approximately 1: 10^{164}. That's a trillion, trillion, trillion, trillion, trillion, trillion, trillion times smaller than the odds of finding a single specified particle among all the possible particles in the universe. It was originally believed that billions of years were available for life to originate by purely natural processes. But we now have fossil evidence of life going back as far as 3.8 billion years. Now, when one reflects that the age of the earth itself is probably 5 to 6 billion years old, that means that the window of opportunity between the time that the earth cooled down and the seas formed on the one hand and the first fossilized life 3.8 billion years ago on the other, is being progressively closed. Sometimes people will say that if the universe is infinite in size, then no matter how improbable the origin of life is, it will originate somewhere by chance. The problem with this objection is that it could be used to explain away any event, no matter how improbable! Rational behavior would become impossible. In fact, on this view, we could never have any evidence that the universe is infinite, because if it were infinite, it would become impossible to assess the probability or improbability of the evidence. So you could not even have any evidence that the universe is infinite if it is infinite. Thus the objection is empirically incoherent and cannot be rationally affirmed.

We don't know what means God used, or whether he used means, to bring about the origin of life. But I think we can say that the scientific evidence is certainly consistent with the origin of life's being, in Francis Crick's words, "a miracle," that is to say, an event which was supernaturally brought about by God.

As for the evolution of biological complexity and human beings, the theist is ready to follow the scientific evidence where it leads. While the evidence for common ancestry can be impressive, the evidence for the efficacy of random mutations and natural selection as the engine of evolutionary development is scanty, to say the least.

It's worth emphasizing just how extraordinary an extrapolation the received theory involves. Typical of the evidence offered on behalf of the efficacy of random mutations and natural selection are: the experience of breeders; the peppered moth experiments; the development of drug resistance by bacteria. Is there any reason to extrapolate from these limited cases to the grand evolutionary story? Many of us probably think that if random mutation and natural selection can explain, for example, the evolution of the horse, then that surely shows the power of these mechanisms. In fact, evolution within a single kind like this is nothing compared to the vast range of life. One might think that if we could show that random mutation and natural selection could explain, say, how a bat and a whale evolved from a common ancestor, that would truly show the power of these mechanisms. Think again.

The following figure shows the major groups or phyla of the animal kingdom. Notice that a bat and a whale are both mammals, which is just one of the subcategories under the chordates. Even the evolution of a bat and a whale from a common ancestor is an utter triviality compared to the vast range of the animal kingdom.

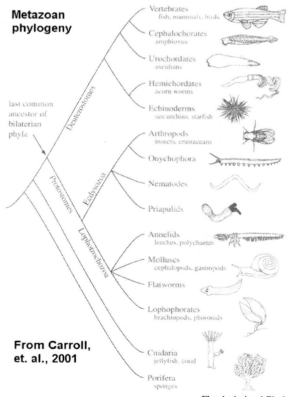

Fig. 1: Animal Phyla

Success in such a case would do nothing to explain, for example, how a bat and a sea urchin evolved from a common ancestor, not to speak of a bat and a sponge. This represents an extrapolation of gargantuan proportions.

Now if this extrapolation takes our breath away, consider the following figure:

Phylogenetic Tree of Life

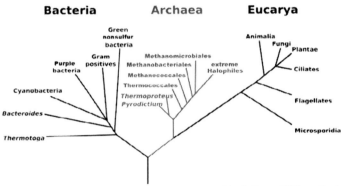

Fig. 2: Tree of Life on Earth

The whole of Fig. 1 is contained on the little twig of the right hand branch of Fig. 2 labeled "Animals." And notice the twig to the right of it: "Plants." The whole of the plant kingdom is contained in that twig. And these are just *two* twigs on the branch of Eucaryotes. There are still the two other domains of the Bacteria and Archaea to be accounted for. The extrapolation of the neo-Darwinian mechanisms from peppered moths and fruit flies and finch beaks to the production and evolution of every living thing is a breathtaking extrapolation of gargantuan, brobdingnagian proportions. We know that in science such extrapolations often fail. So, again, I think we're compelled to ask, what is the evidence for this extraordinary extrapolation?

In their book *The Anthropic Cosmological Principle*, physicists John Barrow and Frank Tipler list ten steps in the course of human evolution, each of which—*each of which*—is so improbable that before it would occur the sun would cease to be a main sequence star and would incinerate the earth. These include things like the development of DNA-based genetic code, the evolution of aerobic respiration, evolution of glucose fermentation to pyruvic acid, the development of an endoskeleton, and so on and so forth. As a result they report, "There has developed a general consensus among evolutionists that the evolution of intelligent life, comparable in information-processing ability to that of Homo sapiens, is so improbable that it is unlikely to have occurred on any other planet in the entire visible universe."

But that raises the question, then, why think that it has evolved on this planet by these neo-Darwinian mechanisms? Indeed, doesn't the evidence suggest just the opposite? A progressive creationist view involving periodic divine causal interventions seems to fit the evidence better than naturalism.

But if there is no conflict between evolutionary biology and generic theism, is there, nonetheless, a conflict between evolutionary biology and Christian, *i.e.*, biblical, theism? That will depend in large measure upon our interpretation of Genesis 1. In order to interpret this passage correctly we have to follow a fundamental hermeneutical principle of interpreting a writing according to the literary genre or type to which it belongs. Considerations of genre are crucial because it would be a mistake to read a literary text literally if the genre of that text isn't the sort that is intended to be taken literally. For example, when the Psalmist says, "Let the trees of the wood clap their hands before the Lord," he's not trying to teach botany! The Psalms belong to the genre of poetry, and it would be a misinterpretation of the text to construe it literally. Another hermeneutical principle that we should follow is to try to determine how the original author and audience would have understood the text. We should examine the text in its own historical context and culture and not attempt to read modern science into the text.

So the question is whether Genesis 1 is of a genre that intends the reader to take it literally. This is doubtful. Whereas 19th-century scholars looked at ancient creation myths as a sort of crude proto-science, contemporary scholars tend more to the view that such stories were taken figuratively, not literally, by the people who told them. When we compare Genesis 1 to Egyptian creation stories, significant similarities and differences emerge. Genesis 1 is an incredibly carefully crafted piece of Hebrew literature. It is not poetry; it is not a hymn; but it is not just straightforward prose either. It is a highly stylized piece of writing exhibiting carefully crafted parallelism—for example, "and God said . . . and God made . . . and it was so." It is a very carefully, stylistically structured chapter that exhibits a great deal of literary polish. It is not simply a scientific report of what happened.

In particular, for example, it is unwarranted to press the Hebrew word *yom* ("day") to mean a literal 24-hour period of time. Note that the phrase "and it was evening and it was morning" is not mentioned with respect to the seventh day. That suggests that the seventh day is intended to be still continuing. The Genesis account itself uses the word *yom* in a metaphorical sense to describe the entire creation week, not just a 24-hour period of time (Gen. 2.4). A problem that has bedeviled literalists from earliest times has been the fact that God doesn't make the sun until the fourth day. But if that is the case, then how could the previous days been 24-hour periods of time, if there wasn't any sun to create solar days?

Furthermore, notice something peculiar when it comes to the third day. The text does *not* say, "And God said, 'Let there be fruit trees bearing fruit after their kinds and vegetation bearing seed after their kinds!', and it was so." Rather it says, "Let the *earth* bring forth fruit trees, *etc*.! And the earth brought forth, *etc*." Now ancient Israelites knew how long it takes, for example, for an apple tree to grow from a tiny sapling to a mature tree, which will then blossom and bear apples. If the author were thinking here of a 24-hour period of time, he would have to be imagining something that would look like time lapse photography, where we see the plants burst out of the ground and grow up into full maturity, and then blossoms pop out, and then the fruit pops out on the branches! I can't persuade myself that the author of Genesis is imagining that kind of process, like a film being run on fast-forward, when he says the earth brought forth vegetation, bearing seed according its kind, and trees bearing fruit according to their kinds. So I think it's very plausible to think that the author is not imagining this happening in a literal 24-hour period of time. The narrative is figurative in nature. All this is said independent of the input of modern science.

Turning to the *human mind*, I believe that we human beings are composite entities composed of soul and body and that these interact in order for us to function properly in this life. Reductive or eliminative forms of materialism are increasingly unpopular. They just don't seem to account for our mental lives because the brain, as a physical substance, has simply physical properties, like a certain volume, a certain mass, a certain density, a location, a shape; but the brain doesn't have mental properties. The brain isn't jubilant, the brain isn't sad, the brain isn't in pain. So the brain alone, as a physical glob of tissue, doesn't have the mental properties that are characteristic of mental states. This has led many thinkers to affirm some sort of non-reductive physicalism—the view that the brain gives rise to epiphenomenal states of awareness like jubilance or sadness or pain. But there isn't any *thing*—there isn't any soul or mind—that is in these states. Rather, the brain is the only thing that really exists.

There are a number of problems with this view that make it improbable. First of all, *it is incompatible with self-identity over time.* Think about it: the brain endures through time and so has diachronic identity; but these states of awareness don't endure from one moment to the next.

This view of the self—the "I"—is rather like the Buddhist view of the self, which holds that the soul or the self is like the flame of a candle. The candle and the wick endure from one moment to the next, but the flame doesn't endure. Every state of the brain at different times has a state of awareness associated with it, but there isn't any enduring self or "I" that endures from one moment to the next. So if you believe that you are the same person who began reading this answer, you ought to reject the non-reductive physicalist view of the self.

Second, *intentional states of consciousness don't seem to make sense on this view*. The property of intentionality is the property of being about something or being of something. For example, I can think *about* my summer vacation or I can think *of* my wife. Physical objects don't have these sorts of properties. The brain is not about something any more than a chair or a table is about something or of something. It is only thoughts which have this kind of aboutness or intentionality. But on this view there is no self—no soul—which has the property of intentionality. Intentionality is just an illusion. But an illusion is an illusion *of* something. So an illusion of intentionality is itself an intentional state. So the view that intentionality is merely an illusion is literally self-refuting and incoherent. If you think, again, that you ever have thoughts about something or of something, you ought to believe in the reality of the soul and reject physicalist views.

Thirdly, *free will seems impossible to reconcile with either reductive or non-reductive physicalism* because on these views there is no causal connection between the states of awareness. The only causality is on the physical level, and that is totally determined by the laws of nature and the initial material conditions. So there just isn't any room for freedom of the will. So on this view, again, free will is an illusion—you never really do anything freely. And that flies in the face of our experience of ourselves as free agents. So if you believe that you ever act freely— for example, in believing in determinism!—that gives you reason to believe in the reality of the soul and to reject these reductive and non-reductive physicalist views.

Finally, *mental causation is incompatible with non-reductive physicalism*. On non-reductive physicalist views, the only arrow of causation is from the brain to epiphenomenal states. The epiphenomenal states themselves are causally impotent. Why? Because there is nothing there—no soul, no mind—that can exert a causal influence on the brain. But I do have a causal influence on my brain states. I can cause my arm to go up by willing to do so. I can do other things through thinking and thereby bring about causal effects.

So it seems to me that we ought to reject these physicalist views in favor of some sort of dualism-interactionism. As Sir John Eccles has put it, the soul uses the brain as an instrument for thought, just as a pianist uses a piano as an instrument to produce music.

Let me say something about challenges to this view emanating from the experiments performed by Benjamin Libet, which suggest that we become conscious of our decisions after making them. His results are exactly what the dualist-interactionist would expect. The soul or the mind doesn't work independently of the brain. So, *of course*, the soul's decisions are not simultaneous with the soul's awareness of those decisions—how could they be? Given the soul's reliance upon the brain and the finite velocity of neurosignals, there would have to be a time lag between the soul's decision and the soul's conscious awareness of that decision. In Libet's experiments, since the neurosignals travel at finite velocities, it takes a little time for the soul's decision to come to conscious awareness. This is exactly what we should expect on a dualist-interactionist view. We are composites of soul and body, and the soul and the body—in particular the brain—work together to think.

Finally, as far as *ethics* is concerned, science can at the very most tell us something about the way in which we have come to hold our moral beliefs, but it says nothing about the objective truth of those beliefs or their ontological foundation. Evolutionary psychology at best tells us how our subjective perception of moral values and duties has evolved. But if moral values are gradually discovered, not invented, then our gradual and fallible apprehension of the moral realm no more undermines the objective reality of that realm than our gradual, fallible perception of the physical world undermines the objectivity of that realm. As a theist I see objective moral values as grounded in God, the paradigm of goodness, and objective moral duties as constituted by His commands.

3. Some theorists maintain that science and religion occupy non-overlapping magisteria—i.e., that science and religion each have a legitimate magisterium, or domain of teaching authority, and these two domains do not overlap. Do you agree?

Obviously not! The circles representing these domains intersect.

4. What do you consider to be your own most important contribution(s) to theorizing about science and religion?

I suppose it would the integration of theology and science to formulate a synoptic worldview, *e.g.*, my work on God and time.

5. What are the most important open questions, problems, or challenges confronting the relationship between science and religion, and what are the prospects for progress?

I think that neuro-science and neuro-theology will be the new frontier. I'm confident that a synoptic understanding will be available in this field as well.

5

William Dembski

William Dembski is an American mathematician and philosopher and one of the world's leading proponents of intelligent design (ID). He is currently Senior Fellow with the Discovery Institute's Center for Science and Culture and Senior Research Scientist with the Evolutionary Informatics Lab. Until the summer of 2013, he was Phillip E. Johnson Research Professor of Culture and Science at Southern Evangelical Seminary in Charlotte, North Carolina, where he helped head its Institute of Scientific Apologetics. His books include *The Design Inference: Eliminating Chance through Small Probabilities* (1998), *Intelligent Design: The Bridge Between Science & Theology* (1999), *No Free Lunch: Why Specified Complexity Cannot be Purchased without Intelligence* (2002), *The Design Revolution: Answering the Toughest Questions about Intelligent Design* (2004), *The End of Christianity: Finding a Good God in an Evil World* (2009), and *The Nature of Nature: Examining the Role of Naturalism in Science* (co-edited with Bruce Gordon).

---

1. What initially drew you to theorizing about science and religion?

I was raised a very nominal Roman Catholic with little interest in organized religion and only too happy to chuck it once I was no longer compelled to attend church. Even after I began to take the Christian faith seriously (at age 18), the relation of science to faith was the least of my concerns (even though I had a father who was a biology professor). For me, coming to faith in Christ was personal and existential—I needed to know that God knew (really knew, from the inside out) the human condition of suffering and could meet us there. Christ in the Incarnation and through the Cross met that need independently of any insights or critiques that science might offer to religion generally or to Christianity in particular.

Nonetheless, once I started trying to make sense of my Christian faith in light of the wider culture, and especially as pertains to the loss of credibility that Christianity had in recent decades suffered in the West (Ross Douthat documents this in *Bad Religion*, but I experienced it first hand), the role of science, and especially its use in buttressing a materialist worldview, came to occupy me. I saw no conflict between Christianity and evolution if evolution were construed merely as the claim that organisms had changed, even drastically, over time. What troubled me was the eradication of teleology from evolution, and the consequent use of evolution as a stalking horse for atheistic materialism.

Indeed, why did Andrew Dixon White open his *History of the Warfare of Science with Theology in Christendom* not with the Galileo episode (Galileo doesn't appear until chapter 2) but with Darwin and evolution? I read significant chunks of that book as a mathematics postdoc at MIT in 1988. Now, one might say that White's views on science and religion were skewed and unduly negative toward religion, reveling in science's role as a nemesis of faith. But shortly thereafter I read Richard Dawkins' *The Blind Watchmaker*, where early in the book he states that Darwin made it possible to be an intellectually fulfilled atheist. Here was a clear trend of using science as a club to beat religion. That trend continues to this day, as when the late Christopher Hitchens writes a book titled *God Is Not Great*, and then devotes a full chapter to touting Darwin and critiquing intelligent design. Why does Hitchens, a "lit guy," need to invoke science—a materialist conception of science that puts Darwin's theory of evolution front and center—to bolster his atheism?

The answer to this question is straightforward: to be an atheist, one needs some account of how we got here that makes no use of God or any real teleology. Darwin, through his theory of natural selection, offered such an account. It's not that natural selection demands that there be no God—God could, as a logical possibility, create through natural selection. But the point is that there is nothing in an evolutionary process, once controlled by natural selection, that requires God or points to him or to any purposiveness in nature. A purely material universe, left to its own devices and without the need of a deity, could, as far as the dominant understanding of evolution is concerned, bring about life (Francisco Ayala, in his writings, has made this point as clear as possible).

Now, if I thought that reason and evidence for this evolutionary view were persuasive, I might have been swept into this materialistic form of thinking, possibly adding a theistic twist to it, as the theistic evolutionists do. But I did not find theories of chemical and biological evolution persuasive. It seemed to me that all the heavy lifting in these theories was assigned to the stochastic properties of matter, and these, it seemed, simply could not do the work expected of them. As it is, I ended up becoming a probabilist (Patrick Billingsley and Leo Kadanoff were my doctoral supervisors at the University of Chicago), and turned that expertise in probability theory into a research project for understanding what in nature is within and outside the reach of chance. Much of my work in intelligent design keys off of this distinction. I regard intelligent design as a scientific program for understanding the activity of teleological causes in nature. Intelligent design therefore has implications for the dialogue between science and religion, implications I have addressed in some of my writings, but implications that are not properly part of intelligent design's scientific program.

2. Do you think science and religion are compatible when it comes to understanding cosmology (the origin of the universe), biology (the origin of life and of the human species), ethics, and/or the human mind (minds, brains, souls, and free will)?

It depends on what you mean by science. The very term science, as we use it these days, is less than 200 years old. *Natural philosophy* is the term that used to denote science, and its history covers more than two millennia. Now there has not been a single natural philosophy, or science, over that time. Just what science or natural philosophy is, is not written in stone. There's been Aristotelian science, in which teleology played a substantive role. There's been Epicurean science, in which atoms were fundamental, though the Epicurean atoms were not wholly deterministic but instead subject to random swerves. There's been Newtonian science, which saw the world as particles subject to purely deterministic forces. There's been the science of the quantum in which stochastic and nonlocal effects play an ineliminable role. There's been science that sees nature as purely material. There's been science that does not prejudge nature, allowing evidence rather than presupposition to decide the nature of nature.

So, in answer to this question, I would say religion is fully compatible with science provided science is not straitjacketed by materialism. For a nonmaterialist science, the universe and life can give evidence of resulting from purpose, ethics can be real rather than a mere social construction, mind can be more than merely patterns of brain activity, and humans can be truly free and not just free in a soft determinist sense where free actions are defined as actions produced without outside coercion (but merely the internal coercion of a determined brain). For a materialist science, which presupposes that matter is all there is, it is no coincidence that cosmology, biology, ethics, psychology, and neuroscience all run into conflict with deeply held religious intuitions.

But what is such a nonmaterialist science? Various proposals are out there on the table. I've got my own through intelligent design. Rupert Sheldrake has his through morphic resonance. Nonmaterialist neuroscientists such as Mario Beauregard find compelling evidence of mind transcending brain. Unfortunately, in this cultural climate, materialists reject nonmaterialist science as "pseudoscience." Such pejoratives are easily tossed around, and yet can be quite effective at shaping public opinion given that materialist science is ascendant in our culture. That ascendancy is, I believe, coming under pressure from nonmaterialists (witness Thomas Nagel's *Mind & Cosmos*), but the distinction between science as such (a dispassionate investigation of nature) versus science as applied materialism (a biased inquiry that shoehorns nature into materialist categories) is one that many mainstream scientists have difficulty drawing. As soon as one puts on materialism's rose-colored glasses, the thought that science may have more to study than matter goes by the board.

3. Some theorists maintain that science and religion occupy non-overlapping magisteria—i.e., that science and religion each have a legitimate magisterium, or domain of teaching authority, and these two domains do not overlap. Do you agree?

I've never been a fan of Stephen Jay Gould's NOMA (non-overlapping magisteria), if for no other reason than that Gould himself could not consistently follow his own advice in keeping science and religion separate. As he set up the problem, science deals with matters of fact, religion with matters of ethics and personal experience. Yet Judaism and Christianity, the religions Gould encountered the most in his lifetime, make claims about history, notably miraculous occurrences, such as Elijah calling down fire from heaven or Jesus resurrecting from the dead. Gould, of course, had to reject such occurrences because these could not be facts given his conception of science and how he saw it as constraining physical reality. And yet, such miraculous events are part and parcel of many people's religious faith. So Gould, to keep his separate magisteria, had to redefine science as a materialist enterprise and religion as a purely personal exercise.

Gould was cagey about admitting that he had in fact tendentiously redefined science and religion to ensure that they would not conflict (a long footnote near the middle of *Rocks of Ages* is the clearest he gets on this topic). Other writers have been more straightforward about the need for religion to get in line with science. For instance, my debate partner Michael Ruse (we co-edited *Debating Design: From Darwin to DNA* for Cambridge University Press a decade back) wrote a book titled *Can a Darwinian Be a Christian?* There he allows that Darwinians can be Christians provided that Christianity can be made compatible with unbreakable natural law. A consequence of this requirement is that the Resurrection of Jesus must be understood not as a real miraculous event. As Ruse puts it, "Even the supreme miracle of the resurrection requires no law-breaking return from the dead. One can think of Jesus in a trance, or more likely that he really was physically dead but that on and from the third day a group of people, hitherto downcast, were filled with great joy and hope." I would much prefer that Gould and Ruse simply say that Christianity is nonsense (like Richard Dawkins) than go through such contortions.

My own view is that everything hangs together with everything else, if not proximately then in a finite series of relations. Accordingly, to isolate science and religion from each other is, in my view, always a vain exercise. Nor do I regard one of these, science or religion, as privileged over the other. I therefore reject the view that "science always knows better," and thus religion must get in line with the most recent pronouncements of science. Likewise, I reject that "religion always knows better," and thus science must bend to ecclesiastical dogma. God has, in my view, revealed himself in both nature, as treated by science, and history, as recorded in sacred texts. Bringing these together coherently is a challenge—neither a simple concordism nor a reflexive trumping of one by the other will do.

For instance, it seems absurd to me for someone to think (as does David Bartholomew) that quantum indeterminacy requires that even God cannot know future chance events. God, who created a world that operates by quantum mechanics, need hardly be bound by its principles. Science, it seems to me, simply cannot speak to divine omniscience one way or another. The Christian theological tradition (process and open theism notwithstanding), by contrast, is clear in affirming full divine omniscience (God knows past, present, and future perfectly). Conversely, it seems absurd to me to insist that the universe is only 6,000 years old because a particular reading of Genesis requires it. Many convergent lines of evidence from different special sciences witness to a much older universe. God's general revelation in nature needs to be taken seriously and respected when that revelation is clear. Ditto for God's special revelation in Scripture.

4. What do you consider to be your own most important contribution(s) to theorizing about science and religion?

In 2005 I started the blog UncommonDescent.com. I no longer run this blog (my friend and colleague Barry Arrington took it over some years back). But on the "about page," which is there to this day, I described the vision of the blog as follows:

> Materialistic ideology has subverted the study of biological and cosmological origins so that the actual content of these sciences has become corrupted. The problem, therefore, is not merely that science is being used illegitimately to promote a materialistic worldview, but that this worldview is actively undermining scientific inquiry, leading to incorrect and unsupported conclusions about biological and cosmological origins. At the same time, intelligent design (ID) offers a promising scientific

> alternative to materialistic theories of biological
> and cosmological evolution—an alternative that is
> finding increasing theoretical and empirical sup-
> port. Hence, ID needs to be vigorously developed
> as a scientific, intellectual, and cultural project.

I give this quote because it encapsulates my approach to the science-religion dialogue. In my view science—an ideologically invested science
where the ideology in question is materialism—has run roughshod over
religion. My contribution to this dialogue, as I see it, is to help stop this
abuse of religion by science. Now there are two ways to stop the abuse.
One is to assert, as Alister McGrath does, that the content of science as
we now have it is just fine and that what's really the problem is people
like Richard Dawkins who illegitimately draw anti-religious implications from science. If you will, McGrath will grant Dawkins his science
but question his metaphysics. The more radical approach, and the one
I and fellow ID theorists adopt, is to challenge Dawkins not just on
his metaphysics but also on his science, arguing, in particular, that key
aspects of his understanding of evolution (such as the creative power of
natural selection) are not based on evidence (despite his protestations
that the evidence is "overwhelming") but follow as logical consequences of his atheistic materialism. His key claims for evolution, therefore,
do not stand up under scrutiny.

Left here, my theorizing about science and religion would be mainly
negative in the sense that I would simply be attacking certain scientific
claims. To be sure, some of my work here is highly critical and negative
in this sense. But it is a truism that deeply controversial issues are not
settled by simply attacking a position. Instead, their resolution depends
on presenting a coherent alternative. My work on intelligent design attempts to present such an alternative, making a positive contribution
to science by clarifying the role of intelligence or teleology in nature
while pointing up the inadequacies of a science wedded to materialism.

As a probabilist, I use probability theory to analyze teleological alternatives to evolution. Some critics of intelligent design have argued that
probability is irrelevant to these discussions, but in doing so they are
either uninformed or disingenuous. Whenever ID-critic Kenneth Mil-
ler, for instance, cites experimental evidence for the power of natural
selection, he appeals to some experimental set-up in which selection
pressure brings about, with high probability, some biological structure/
function previously lacking. But if high probability provides confirming
evidence for Darwinism, why can't low probability provide disconfir-
ming evidence? Parity of reasoning demands that if probabilities can

support Darwinism, then they can also put it in empirical harm's way.

When I began to examine the probabilistic hurdles facing Darwinian natural selection, I didn't see probabilistic methods as making a positive case for teleology so much as making a negative case against materialism. Materialistic processes alone, in the absence of real teleology, give an incomplete account of nature. The case for this negative claim is strong, a point recently drilled home by Thomas Nagel in *Mind & Cosmos* (Nagel even refers to materialism as requiring "improbable flukes"). But probability, in conjunction with pattern, can also be used to make a positive case for intelligence or real teleology in nature. When events of small probability are suitably patterned (or "specified"), the convergence of probability and pattern is doing more than underscore the incompleteness of materialistic processes—this convergence is also pointing to real teleology in nature.

I show how *specified complexity* provides a reliable method for detecting intelligence/teleology in my monograph *The Design Inference* (Cambridge University Press, 1998) and then further clarify that method and apply it to biological systems in subsequent work (e.g., *The Design of Life*, co-authored with Jonathan Wells). My research over the last seven years with the Evolutionary Informatics Lab (EvoInfo.org— see its publications page) takes specified complexity and translates it into an information-theoretic apparatus for measuring and tracking the information needed to build biological systems. Moreover, in my forthcoming book with Ashgate titled *Being as Communion: A Metaphysics of Information*, I show how information can be conceived as the primary stuff of reality and how matter itself can be viewed as a type of information.

All of this work, varied though it is, constitutes a single trajectory for trying to make sense of teleology in nature, and together represents my contribution to the science and religion dialogue.

5. What are the most important open questions, problems, or challenges confronting the relationship between science and religion, and what are the prospects for progress?

The information-theoretic underpinnings of physical reality seem to me to constitute the most important and intriguing area of research for the science-religion dialogue. John Wheeler, Paul Davies, and many other high-profile scientists have pondered the role of information in nature, suggesting that it is the most basic object in nature and of scientific inquiry. If it is, then not only is materialism dead but the door is opened to information, consciousness, and mind being, as John Horgan muses, "not an accidental by-product of the physical realm but ... in some sense the primary purpose of reality." Research at the Evolutionary Informatics Lab (EvoInfo.org) gives me confidence to think that we're making excellent progress toward such an "informational realism."

In answering these five questions, I've been hammering on materialism. I've done so because, in the dialogue between science and religion, materialism has subverted the content of science and therewith radically undermined religion. If matter is all there is and if the principles by which it operates are at base nothing but mindless forces of attraction, repulsion, and the like, then religion becomes an epiphenomenon of matter and loses any purchase on telling us the ultimate nature of the world. Religion has traditionally claimed to give us such insight into the world, but a materialist science ensures that all of religion's high ambitions are merely pretense and grandiosity. I therefore see as a vital part of the science-religion dialogue developing a thorough critique of materialism's shortcomings and making sure that critique is widely disseminated in the academy as well as to the wider culture. Unseating materialism in this way remains a work in progress, with materialism still dominant among the West's intellectual elites, though the arguments against it grow continually stronger.

So much of the relationship between science and religion in the 1980s and 1990s, when I got into this business, was concerned with reconceptualizing religion in light of the challenges that a materialistic science threw at religion (especially non-teleological evolutionary theories—cosmological, chemical, and biological). The need to dismiss or radically reconceive religious ideas such as the supernatural, divine transcendence, divine action, divine power and knowledge, teleology in nature, ethical realism, etc. has all flowed, as I see it, from this commitment to a materialist science. Over and over I saw a materialist science used as a tool to gut religion of the things that ordinary people find important and meaningful. Over and over I saw smug intellectuals telling us what we could no longer believe about religion because science has shown us how wrong the best religious thinkers and prophets of the past have been.

Fortunately, I see this destructive approach to the science-religion dialogue as waning. Instead, I see more constructive work that explores resonances between science and religion without trying to invalidate one or the other (the Templeton Foundation deserves some credit here for encouraging this sort of work). For instance, does following traditional religious morality lead to healthier brain states (as gauged, perhaps, by a neuroscientist's fMRI)? Is Darwinian survival and reproduction inherently incapable of explaining certain of our moral impulses (such as real as opposed to reciprocal altruism), and are these impulses instead better explained within a broader religious framework? Do spiritual experiences require complex interactions of brain activity that cannot be dismissed as an overly excited "God spot" in the brain (neuroscientist Mario Beauregard's work on the spiritual experiences of Catholic nuns suggests that a reductive materialism cannot explain away such experiences).

Although the science-religion dialogue addresses many deep and thorny conceptual issues, it also faces a practical challenge, namely, how to get various parties in that dialogue to stop attacking each other and to find opportunities for free and fruitful exchange of ideas. Some factions in this debate are only too ready to demonize other factions. In my own experience, I found fundamentalist Christians and fundamentalist materialists both trying to get me fired from various academic positions I've held because I did not adhere to their views of how science and religion should properly be related. Fundamentalism is, in my view, the worst enemy of the science-religion dialogue, and most fundamentalists don't know that that's what they are. The science-religion dialogue needs to be a lively conversation in which all who are willing to inform themselves and act civilly should be welcome. Fortunately, I see more and more people with open minds and good wills ready to take part in that conversation.

6

Daniel C. Dennett

Daniel C. Dennett is a world-renowned American philosopher and cognitive scientist. He is currently University Professor and Austin B. Fletcher Professor of Philosophy, and Co-Director of the Center for Cognitive Studies at Tufts University. His research centers on the philosophy of mind, free will, and philosophy of biology. He is also well known for his atheism and is often referred to as one of the "Four Horsemen of New Atheism" (along with Richard Dawkins, Sam Harris, and the late Christopher Hitchens). He is the author of several influential books including *Content and Consciousness* (1969), *Brainstorms* (1978), *Elbow Room: The Varieties of Free Will Worth Wanting* (1984), *The Intentional Stance* (1987), *Consciousness Explained* (1991), *Darwin's Dangerous Idea* (1995), *Kinds of Minds* (1996), *Freedom Evolves* (2003), *Breaking the Spell: Religion as a Natural Phenomenon* (2006), *Science and Religion: Are They Compatible?* (co-authored with Alvin Plantinga) (2011), and *Intuition Pumps and Other Tools for Thinking* (2013). In 2004 the American Humanist Association named him Humanist of the Year, in 2010 he was named to the Freedom from Religion Foundation's Honorary Board of distinguished achievers, and in 2012 he was awarded the Erasmus Prize.

---

1. What initially drew you to theorizing about science and religion?

Early in the 21st century I had become concerned, as many others had, that the Religious Right had overstepped the bounds of reasonable activism and proselytizing, and were threatening to push America towards a theocracy. (It is hard to remember how aggressive and threatening these factions were in those days, but they had both the media and the politicians cowed into submission, or so it seemed to many of us.) I decided the day of the diplomatically silent atheist was over; it was time to come out in public and assert my point of view in hopes of encouraging others to do likewise. When I published an op/ed piece about the attempt to re-label atheists, freethinkers, humanists and agnostics as "brights," (titled, by the NYTimes: "The Bright Stuff") in July of 2003, I received hundreds of messages urging me to write more about atheism and religion. I had been working on cultural evolution and its biological background, and a natural application of the insights I'd been garnering from that work was an assessment of religions as natural phenomena. I set aside my main research interests for three years and devoted myself full time to researching and writing *Breaking the Spell*.

2. Do you think science and religion are compatible when it comes to understanding cosmology (the origin of the universe), biology (the origin of life and of the human species), ethics, and/or the human mind (minds, brains, souls, and free will)?

No, science and religion are not compatible on these topics, except on those rare occasions when religious texts happen to express truths inadvertently (ancient people weren't wrong about everything, after all). The creation myths of religion are interesting pre-scientific fictions, devised by people who did not yet have the perspectives or thinking tools of science to aid them in their queries, and they reveal a lot about both the powers and limitations of human imagination, but they have nothing else to add to our understanding of how the universe came to be as it is and what is to become of it. The portrayals of people we get in religious narratives are often illuminating, demonstrating at the very least that ancient people had much the same psychology as we have, and responded to life's opportunities and challenges as we do today. Among them are many cautionary tales that, by being "universally" known, help to stabilize our mutual expectations about how people treat each other—but we get just as many, and often better, narratives from Aesop and Homer and the other great storytellers over the millennia. Other Biblical tales are more problematic to today's audiences, full of breathtaking cruelty and baffling excesses of pride or hatred, and their enigmatic quality probably explains their staying power; we try, unsuccessfully, to conjure a settled context in which we can see why folks found this tale worth repeating, and our unsatisfied curiosity propels the story down through the generations. We have moved so far beyond Biblical morality, especially Old Testament morality, that pastors and Sunday School teachers have to perform heroic framing exercises to make large parts of the Bible decent fare for young people, and many verses are either consigned to oblivion or reserved for mature audiences that can somehow rationalize the presence of such indefensible moral judgments in their holy scripture.

3. Some theorists maintain that science and religion occupy non-overlapping magisteria—i.e., that science and religion each have a legitimate magisterium, or domain of teaching authority, and these two domains do not overlap. Do you agree?

Stephen Jay Gould's idea of NOMA was a well-intentioned try, but utterly hopeless as a reconciler of science and religion. It stripped religions—rightly—of all pretense to be a source of factual truth, but this went too far for all but the most hyper-liberal and "sophisticated" religionists, while wrongly stripping secular investigations in the humanities, primarily philosophy, of any authority in the domain of ethics and the meaning of life. How could Gould, a Harvard professor, declare that his Harvard colleagues John Rawls, Amartya Sen, and Robert Nozick—to name just three of the panoply of brilliant non-religious, secular thinkers—were poaching on the magisteria of religion? Religion is no more authoritative about ethics than it is about science. The advances in moral perspective of the last two millennia—the abolition of slavery, the prohibition of cruel and unusual punishments, the tempering of judgments about all manner of "sins"—have been a series of triumphs of secular reasoning over religious conservatism. Gould's proposal satisfied nobody: religious leaders were not ready to consign all their scriptural narratives to the disarmed categories of myth and metaphor, and secular theorists in a variety of fields—philosophy, economics, political science, literary theory, anthropology—were unwilling to accept that they could offer no solid guidance on ethical questions, whereas religious thinkers could.

4. What do you consider to be your own most important contribution(s) to theorizing about science and religion?

I think my analysis of belief in belief is perhaps my best contribution. I pointed out that belief in God actually takes a back seat to belief in belief in God: more people believe in belief in God than actually believe in God. That is, they think believing in God is a good thing, even if they can't muster the conviction themselves. In fact it is very difficult for anybody to show evidence that they believe in God—beyond the evidence of their avowals, which attest to their belief in belief in God, but are strictly neutral as evidence of actual belief in God. The second-order belief has been the source of a mighty river of disingenuousness: politicians avow their belief in God in order to be elected but few would take that avowal as convincing evidence of belief in God, since it is so obvious to everybody that there are deep motivations for saying this whether or not it is sincere. To take a dramatic case, there is no good evidence that the pope believes in God, especially since Catholic doctrine enjoins all good Catholics to profess a belief in God whether or not they actually believe. Under the circumstances, what could a pope do to convince his audience that he really was a believer? Since we know that many do not (any longer) believe in God, the burden of proof lies heavily on those who profess, and there is no obvious way for them to discharge it.

5. What are the most important open questions, problems, or challenges confronting the relationship between science and religion, and what are the prospects for progress?

The main pressing issue is whether religions will continue to demand special treatment on this issue, when there are no grounds for granting it. Religions that can welcome intense scientific scrutiny of their histories, their practices, their creeds and commandments will deserve to survive; the rest, if they wither away under the harsh light, will deserve their fate.

7

George F. R. Ellis

George F. R. Ellis is a theoretical cosmologist and Emeritus Distinguished Professor of Complex Systems in the Department of Mathematics and Applies Mathematics at the University of Cape Town in South Africa. He co-authored *The Large Scale Structure of Space-Time* (1973) with Stephen Hawking, and is considered one of the world's leading theorists in cosmology. He is a Fellow of the Royal Society (FRS), Fellow and past President of the Royal Society of South Africa, past President of the International Society for General Relativity and Gravitation, and past President of the International Society for Science and Religion. He has an A-rating from the National Research Foundation (NRF) and has received numerous awards and distinctions. He is the recipient of the Herschel Medal (RSSA), the South African Institute of Physics De Beers Gold Medal, the South African Association for the Advancement of Science Gold Medal, the South African Mathematical Society Gold Medal, and the Academy of Science of South Africa Science-for-Society Gold Medal. In 1999 Ellis was awarded the Order of the Star of South Africa by President Nelson Mandela for his outspoken opposition to apartheid, and in 2006 President Thabo Mbeki conferred the Order of Mapungubwe on Ellis. Ellis is also the recipient of the Templeton Prize (2004), presented by Prince Philip at Buckingham Palace, for "Progress Towards Research or Discoveries about Spiritual Realities."

---

1. What initially drew you to theorizing about science and religion?

I initially started thinking about these themes through reading *The Nature of the Physical World* by Arthur Stanley Eddington (A. S. Eddington 1928). But this was not an active involvement. I then collaborated on technical papers in cosmology with William Stoeger, a Jesuit scientist attached to the Vatican Observatory. Our discussions extended over time to philosophical issues underlying cosmology, a long standing interest of mine (G. F. R. Ellis 2006). He then invited me to an integrative series of workshops about the interplay between science and religion run by the Vatican Observatory (Castel Gandolfo) in conjunction with the Centre for Theology and Natural Sciences (Berkeley). Through this I learnt much from very knowledgeable and intelligent colleagues, and then wrote a series of articles on the subject and co-authored a book on it with Nancey Murphy (N. C. Murphy and G. F. R. Ellis 1995).

2. Do you think science and religion are compatible when it comes to understanding cosmology (the origin of the universe), biology (the origin of life and of the human species), ethics, and/or the human mind (minds, brains, souls, and free will)?

They are compatible in all these cases, provided one deals with the mature expressions of religion and not the fundamentalist religious wings (whose ideas are anti-scientific). The reason is that they investigate quite different aspects of the universe and humanity by quite different methods. Science deals with testable repeatable universal patterns of behavior of material things; religion deals with issues of meaning, ethics, even to some extent aesthetics. While some aspects of these issues can be explored by scientific investigation, their core nature cannot, because there are no repeatable experiments that can establish their nature. In any case the kinds of abstractions used in scientific modelling are inappropriate to this domain, which deals with unique events and with higher level patterns of meaning that are not scientifically determinable. They involve philosophical investigation and thought, where repeatable laboratory experiments are simply not the appropriate method of investigation. In essence: science deals with "how" questions and mechanisms, religion with "why" questions and meaning. Neither can answer the issues dealt with by the other.

There are however some areas of interaction and tension. As regards the specific questions posed:

- *Cosmology (the origin of the universe).* There are a series of issues here.

    1) **Physical laws.** The nature of what exists, in the sense of what determines the types of entities that exist and the laws that govern their behaviour. For example, what determines the nature of the particles and fields that occur, and the laws of physics that govern their behaviour? More generally, what kinds of causation are there in the universe?

    2) **Cosmological conditions.** The specific initial conditions for particles and fields that determined the unique nature of the one particular universe that actually came into existence.

    3) **Specific Outcomes.** Given 1. and 2., what is the outcome that will occur? How do the laws and initial conditions lead to what we see around us?

The domain of science is the determining of the nature of entities that exist and the laws at work, and then answering the third question, which is a how question. Despite some attempts to claim it can do so, science cannot answer why items 1. and 2. have the nature they do, because these are "why" questions. Those attempts extend science beyond its proper domain of application; they do so by making untestable extrapolations of established science into domains where it may or may not apply. The proposed mechanisms or laws claimed to lead to the existence of the universe certainly are not testable: they are hypothetical speculations. Belief in their validity is faith, not science. Certainly one can make scientifically based speculations about these issues, but then they must be clearly labelled as what they are: speculations.

And if one could succeed in scientifically determining these issues, that would not succeed in answering any fundamental questions. The same "why" questions would then occur in relation to whatever meta-mechanism is proposed to answer cosmological issues. For example the claimed solution of anthropic issues by invoking a multiverse raises all the same questions again as regards the alleged multiverse: Why does it have a nature that allows life to come into existence? What determined its laws of behaviour, and its initial conditions? Finally of course science cannot answer why anything exists—why there was any creation process at all.

A key issue is what types of causation underlie existence. The usual scientific project is an attempt to implement as far as possible the idea of a combination of chance and necessity underlying existence and causation, but it necessarily has to be incomplete. Inter alia, it is clear that goal-seeking and purpose do indeed exist as causal categories in the universe—they underlie all biology and human life. What has to be explained includes the following:

- *Where do the very causal categories of chance, necessity, and purpose come from?*

- *How do these concepts arise and have effect, and what underlying deeper ontological entities or causation do they represent?*

- *How can they even be relevant if there is no ontological referent— some kinds of possibility spaces—that makes the dichotomy between them a meaningful issue?*

These issues surely cannot be explained by referring to necessity: for inter alia it is the very category of necessity that has to be explained (G. F. R. Ellis, 2011).

The deep question at issue is whether there is any kind of purpose or meaning underlying the universe, or it is just something that happened purposelessly by happenstance? If you want to seriously investigate that question, you need to take into account data that relates to purpose and meaning as well as data that relates to physics and astronomy. If you fail to take into account this broader band of data, including philosophy, literature, history, aesthetics, and ethics,[1] of course you will conclude the universe is meaningless. That conclusion is inevitable if you neglect to take into account the relevant data. You are then choosing to ignore fundamentally important aspects of human life, which certainly do exist, as you formulate your world view (G. F. R. Ellis 2008). The fact that science cannot deal with them does not mean they are irrelevant to understanding the nature and origin of the universe. Religion can claim to provide answers in these domains, in a way that can be seen as making sense. Such claims are no more fantastic than many claims that at present pass as part of theoretical physics (M. Wertheim 2013). (where in many cases the need for experimental proof has been largely abandoned) (G. F. R. Ellis 2013). They are of course not provable: belief in them is a matter of faith. There is nothing wrong with this: faith is a key element of all human life (G. F. R. Ellis 2007). It is part of a holistic human existence.

Conflict arises if either religion tries to answer the "how" questions about cosmology, or science tries to answer the "why" questions about existence that philosophy and religion try to deal with. If they each stick to their own domains, there is no conflict.

- *Biology (the origin of life and of the human species)*: Mature current religious world views have no problem with the idea of Darwinian natural selection and the emergence of life on the basis of physical and chemical principles. These are mechanisms, that is, they are answers to "how" questions. The deep underlying issue is why the universe, through its physical laws and initial conditions, has such a fine-tuned nature as to allow life to come into existence, when such an anthropically favourable nature is extremely improbable (M. J. Rees 1999). Attempts at a purely scientific explanation involve invoking a multiverse in which some pocket universes favour life.

    There are two issues here. First, because this proposal is not scientifically testable apart from some consistency conditions, belief in the multiverse is just that—it's an item of faith (G. F. R. Ellis 2011). There is nothing wrong with this as a philosophical speculation, but it is not proven science. Secondly as mentioned above, this proposal in any case does not solve any fundamental issue. It does not explain why the alleged multiverse has a nature that is favourable to life.

---

[1] P. W. Atkins makes an explicit statement that all these endeavours are of no value in his essay "The limitless power of science" in *Nature's imagination: The frontiers of scientific vision*, Ed. J. Cornwell. Oxford: Oxford University Press, (1995), pp.122-132.

Thus, as has been known for at least a hundred years, there is no conflict between the scientific explanations of evolutionary theory and mature religious world views. Where there can be conflict is over claims about how conditions came about that enabled evolution to ever take place. The problem arises when scientists claim that the hypothesis that a multiverse exists and solves the issue, is an established physics result. It is not. This is a faith based claim.

- *Ethics*: Various efforts have been made to provide a scientific origin for ethics. This is a category mistake, as pointed out *inter alia* by Massimo Pigliucci (M. Pigliucci 2013). These efforts always smuggle in by the back door some highly debatable assumption about what the nature of a good life is; they have to do so, in order to attempt to relate to moral questions. But that is precisely the question at issue, which has been the topic of investigation in over 2,000 years of moral philosophy. It is just delusional for scientists to believe they can solve this fundamental issue, which is a core question in religious life, as a byproduct of scientific investigation. Science simply does not have the capacity to resolve this issue.

   There are no scientific experiments that can determine what is good or bad, and no scientific theory that can determine what is a moral choice of action as regards Iran, Iraq, or Israel, or even global warming. Science has nothing whatever to say about whether existence of polar bears, or rhinos, or human beings, is either good or bad. It can however help determine for how long they are likely to continue in existence.

- *The human mind (minds, brains, souls, and free will)?* There are various aspects to this.

1. **The great gap.** We have no idea how consciousness arises, despite some extravagant claims that consciousness has been explained or that there is no hard problem of consciousness. Despite much investigation of the neural correlates of consciousness, we do not even know how to ask the right questions about how qualia arise. There is not even a beginning of an approach. One should beware of neuroscientists or philosophers claiming much more than has been proven or understood.

2. **Reduction of the mind and denial of free will.** There have been many attempts to claim that because the brain is nothing but a machine/computer, we have no free will, consciousness is an illusion, neuroscience can explain ethics, aesthetics, and religion with no remainder, and so on. For example Semir Zeki claims:

> "It is only by understanding the neural laws that
> dictate human activity in all spheres—in law, mo-
> rality, religion and even economics and politics, no
> less than in art—that we can ever hope to achieve a
> more proper understanding of the nature of man."
> (P. Ball 2013)

That's the reductionist dream—law, morality, religion economics, politics, and art are reducible to neural laws. The implication is they have no validity in their own right. If this were the case, it would indeed be disastrous for religion. But this is not just an attack on religion: it is an attack on our humanity, a denial of our standing as human beings—even though this will be vigorously denied by those putting forward these views. As explained carefully by Merlin Donald in his book *A Mind So Rare*, these views undermine our humanity in a fundamental way.

> "Hardliners, led by a vanguard of rather voluble
> philosophers, believe not merely that consciousness
> is limited, as experimentalists have been saying for
> years, but that it plays no significant role in human
> cognition. They believe that we think, speak, and
> remember entirely outside its influence.... The prac-
> tical consequences of this deterministic crusade are
> terrible indeed. There is no sound biological or
> ideological basis for selfhood, willpower, freedom,
> or responsibility. The notion of the conscious life
> as a vacuum leaves us with an idea of the self that
> is arbitrary, relative, and much worse, totally empty
> because it is not really a conscious self, at least not
> in any important way." (M. Donald 2002)

But these views do not necessarily derive from neuroscience or psychology, because they are based in a bottom-up reductionist view that leaves out the key feature of top down causation in the hierarchy of complexity, which completely changes the situation (D. T. Campbell 1974). One cannot understand developmental biology, (S. Gilbert and D. Epel 2008) physiology, (D. Noble 2008) or the functioning of the mind without taking topdown causation into account. Top-down causation is key for example in perception, attention, goal directed action, and the entire domain of social neuroscience. This allows genuine emergence where our minds have causal powers over physical events: thoughts, social constructions, and emotions influence happenings at many levels in the hierarchy of complexity.

The reductionist perspective explains part of what is going on. To claim it explains all is simply wrong, as a holistic view of brain function will confirm. This neuro-scientific imperialism greatly overstates what neuroscience and brain imaging can achieve, and should be treated with great skepticism.

3. **The fallacy of evolutionary explanation of religion, values, and all social behaviour.** It is often proposed that once you have an evolutionary psychology explanation of something, then this is a complete explanation: it has been fully explained away. If I give an evolutionary psychology explanation of religion, does that make religion false? No. *Everything* the brain does has an evolutionary explanation: but that is only part of the story; and it is irrelevant to the truth value of any belief system. One can equally give an evolutionary explanation for astronomy or astrology, for fairies or for evolutionary psychology, for physics and religion. This has nothing whatever to do with whether any of them are true or not. An evolutionary psychology explanation for any human activity or belief is necessarily always a partial and incomplete explanation, and its existence is irrelevant to the truth of the belief.

4. **Mind and brain: is there a soul?** One of the intriguing aspects of present day views of the brain is the use of the digital computer as a metaphor. Now this is of course a very partial view, nevertheless (remembering that one can simulate any neural network on a digital computer) it does apply to some aspects of how the brain functions. The key point then is that a digital computer is a dualist machine: the hardware does nothing without the software, which is a non-physical entity that inhabits the hardware and controls what happens via its algorithms—which again are non-physical entities. Hence the computer metaphor strongly supports the unfashionable idea of dualism of mind and brain. Taking emotional aspects into account in addition, the concept of a soul is as valid as any other concept to describe the holistic aspects of the way the mind inhabits the body.

Science need not dehumanise us, or deny deep aspects of meaning. More human views are tenable that fully take science seriously, but that do not accept that it is the only route to truth and meaning. The humanities and philosophy do not have to be dictated to by science: one can explore each of them in the rigorous way appropriate to that domain of enquiry, acknowledging the strengths and limits of each of these avenues towards understanding the universe in which we live (J. Kagan 2009). In that context, science can crucially inform philosophy and the humanities, but not supplant them. Philosophy and religion, by contrast to extravagant scientism, affirms human nature and its value: the worth of the individual. Humans are more than machines.

Overall comments:

1. *The Nature of the debate*. Rejecting the possibility of compatibilism because some religious people are fundamentalist and anti-science does not treat the underlying issues seriously. It is like denying the validity of astronomy because of the existence of astrology. There are many deep thinkers who are top class scientists or philosophers as well as being religious, including Nobel Prize winners. Classifying anti-religious people as "Brights"[2] of course implies that all religious people are dim. It manifests a low level and intolerant approach to academic life and to intellectual discussion; see Alistair McGrath's analyses (A. McGrath 2011). of the writings of the self-styed "rationalists," who when their ideas are challenged in a rational manner often resort to insult and personal attack in a highly emotional and irrational fashion.[3] If atheists want to intellectually bolster their case, they need a more sophisticated level of argument that treats their opponents with respect and is able to distinguish between Muslim fundamentalists, Quakers at silent worship, parish priests in rural England, and Martin Luther King, instead of lumping them all together as equivalent when they vehemently proclaim that all religion is evil.[4]

2. *Fundamentalism*. The essential nature of fundamentalism is when a partial truth is proclaimed as the whole truth, discounting all other more complex possibilities. Only one viewpoint is allowed

---

[2] The Brights' Net: "A bright is a person who has a naturalistic worldview. A bright's worldview is free of supernatural and mystical elements." See http://www.the-brights.net/.

[3] An example of the kind of vituperative personal attack considered acceptable by some scientists and their followers as an alternative to rational debate is here: http://whyevolutionistrue.wordpress.com/2013/08/25/famous-physiologist-embarrasses-himself-by-claiming-that-the-modern-theory-of-evolution-is-in-tatters/. A complete answer to the scientific issues raised in that blog is here: http://www.musicoflife.co.uk/pdfs/Answers-new1.pdf.

[4] The allegedly scientific methodology used to "prove" this statement ignores all data about good resulting from religion, as well as all data on good and bad outcomes of atheist movements and regimes. It thus omits the relevant comparisons, and so assumes the desired result before it begins. Incidentally, it also assumes moral realism as a basis for its conclusions.

on any issue, all others are false, and of course we ourselves with our unique insights just happen to be the only people with sole access to the truth. The scientific fundamentalists are just one more example of this kind of view (G. F. R Ellis 2010). The scientism of the New Atheism detracts from its intellectual status (Massimo Pigliucci 2013).

3. Some theorists maintain that science and religion occupy non-overlapping magisteria—i.e., that science and religion each have a legitimate magisterium, or domain of teaching authority, and these two domains do not overlap. Do you agree?

Essentially yes, see previous answer. There is a major problem not only with religious fundamentalists, who falsely claim religion can deal with scientific issues, but also with scientific fundamentalists, who falsely claim science can deal with issues of meaning and ethics.

I believe, in accordance with what Richard Dawkins claims, that a key difference is that science deals with issues that can be observationally or experimentally tested, whereas metaphysics and religion deal with issues that cannot be so tested, and hence are matters of belief or faith. It is a real pity scientists do not stick to the boundaries of what science can actually deal with, according to this standpoint.

Examples:

- Science per se cannot determine how the universe *started*, even though we can make scientifically based theories as to how this may have happened (based in an assumed pre-existing massive machineries of quantum field theories, Lagrangians, symmetry groups, variational principles, etc. etc.; which apparently are supposed to have existed in some Platonic space before time or space existed);

- Science cannot *prove* a multiverse exists, even though it can give scientifically based arguments as to why this may possibly be the case; however one dresses them up, they are in the end philosophical arguments. It is highly ironic when scientists who make a great fuss about how science is about testable theories and so immensely superior to faith, then go on to proclaim the multiverse is the definitive answer to questions of existence.

- Science can say nothing about what is right and what is wrong: there is no *experiment* that can determine what categories of behaviour are good and what are bad. Defining what is good or bad is a philosophical endeavor.

- Science can neither prove nor disprove the existence of God. There is simply no relevant repeatable laboratory experiment or associated data that bears on the issue, which is philosophical territory (P. Russell 2013).

Science is badly served when scientists attempt to take over these territories in a futile attempt at scientific imperialism that rejects philosophy as bunk, while themselves indulging in low level philosophy. Simple minded philosophy is still philosophy; it is just not well-informed.

4. What do you consider to be your own most important contribution(s) to theorizing about science and religion?

- Pointing out that if you want to make a theory of everything that includes issues of meaning and ethics, you'd better take in to account the full range of evidence related to issues of ethics and meaning, and not try to base your theories exclusively on the restricted domain of evidence that can be provided by using telescopes, microscopes, particle colliders, and laboratory experiments. They can't see good and evil, beauty and ugliness, love and hate, or many other things that are deeply important in human life. If you base your world view exclusively on what can be proved by these experiments alone, of course you'll end up deciding the universe is meaningless. You'll have excluded meaning from the equation before you even started.

- Arguing for a kenotic (self-sacrificial) based moral realism as a key feature of the way things are. This nature of deep morality is profoundly paradoxical—you get what you most need by giving it up. This sacrificial nature of morality is discovered by the spiritual wings of all the major religious traditions. It is a key aspect of a good life.

- Pursuing in depth the concept of top down causation and how it works.

- Developing a holistic view of the way the different sciences interact with each other and relate to humanistic and philosophical views of how they all fit together, in a way that takes fully into account both current science and the full depths of humanity as seen in literature, philosophy, and the various religious traditions.

5. What are the most important open questions, problems, or challenges confronting the relationship between science and religion, and what are the prospects for progress?

The most important current issue is to do with the origin of ethics and morality, their nature, and how they are related to each other, as indicated above. The key requirement is a viewpoint that takes ethics seriously in its own terms, as it is indeed a matter of life and death in the real world. Science can help understand neural influences as well as evolutionary influences on individual and social understandings of morality. That is not at all the same thing as characterizing what is actually right and wrong, and what it is that underlies this characterization. If scientists take the limits of their subject and its approach to evidence seriously, they can help in this endeavor. If they claim it can be answered by science alone, that will be a setback and a serious disservice to humanity. You may or may not involve religion, but you must involve metaphysics and moral philosophy in the studies needed.

Progress is indeed possible. It must take into account the nature of the possibility spaces that underlie the way we are able to act and think, and consider the reasons why these possibility spaces (G. F. R. Ellis 2004)—which in some sense underlie the nature of existence in the universe—have the form they do, and in particular why they allow ethical and aesthetic thought as well as logic, mathematics, and physics. They shape the form of possible existence.

- *How and why do they exist, with the specific nature they have? Particularly,*

- *Why is any form of morality whatever possible?*

These are the kinds of deep issues that science cannot begin to address. It assumes existence of these possibilities before it starts.

References

J. Baggott (2013), *Farewell to reality: How fairytale physics betrays the search for scientific truth* (Constable).

P. Ball (2013), "Neuroaesthetics Is Killing Your Soul". *Nature*: 22 March 2013.

J. T. Cacioppo, G. G. Berntson, J. F. Sheridan, and M. K. McClintock (2000). "Multilevel Integrative Analyses of Human Behavior: Social Neuroscience and the Complementing Nature of Social and Biological Approaches" *Psychological Bulletin 126*: 829-843.

D. T. Campbell (1974), "Downward Causation". In *Studies in the philosophy of biology: Reduction and related problems*, F. J. Ayala, and

T. Dobhzansky (eds). (University of California Press, Berkeley)

J. T. Coull and A. C. Nobre (1998), "Where and When to Pay Attention: The Neural Systems for Directing Attention to Spatial Locations and to Time Intervals as Revealed Both PET and fMRI," *The Journal of Neuroscience 18*: 7426-7435.

M. de Haan and M. R. Gunnar (2009) *Handbook of developmental social neuroscience* (The Guilford Press).

M. Donald (2002), *A mind so rare: The evolution of human consciousness* (W. W. Norton).

A. S. Eddington (1928), *The nature of the physical world* (Cambridge University Press).

G. F. R. Ellis (2004), "True Complexity and its Associated Ontology." In *Science and ultimate reality: Quantum theory, cosmology and complexity*. Ed. J. D. Barrow, P. C.W. Davies, and C. L. Harper, Jr. (Cambridge University Press): 607-636.

G. F. R. Ellis (2005), "Physics, Complexity, and Causality" *Nature* 435: 743.

G. F. R. Ellis (2006), "Issues in the Philosophy of Cosmology." In *Handbook in Philosophy of Physics*, Ed. J. Butterfield and J. Earman (Elsevier, 2006), 1183-1285. http://arxiv.org/abs/astro-ph/0602280

G. F. R. Ellis (2007), "The Myth of a Purely Rational Life" *Theology and Science 5*: 87-100 http://www.mth.uct.ac.za/~ellis/Rational_Life.pdf.

G. F. R. Ellis (2008), Backhouse lecture: http://www.mth.uct.ac.za/~ellis/Backhouse_Lecture_rev2.pdf

G. F. R. Ellis (2010), "Fundamentalism in Science, Theology, and the Academy." In *Human identity at the intersection of science, technology, and religion*, Ed. N. Murphy and C. C. Knight (Ashgate), 57-76(http://www.mth.uct.ac.za/~ellis/Fundamentalism.pdf).

G. F. R. Ellis, (2011). "Why Are the Laws of Nature as They Are? What Underlies Their Existence?" In *The astronomy revolution: 400 years of exploring the cosmos* (Taylor & Francis), 387.

G. F. R. Ellis (2011), "Does the Multiverse Really Exist? *Scientific American* 305:380-430.

G.F.R.Ellis(2012),"On the Philosophy of Cosmology" Studies in History and Philosophy of Modern Physics, http://www.mth.uct.ac.za/~ellis/philcosm_18_04_2012.pdf.

G. F. R. Ellis (2012) "Top-down Causation and Emer-gence: Some Comments on Mechanisms," Journ Roy Soc Interface Focus 2: 126-140

G. F. R. Ellis (2013), "Theories Beyond Testability" *Science* 342:934-935.

G. F. R. Ellis (2013), "Multiverses, Science, and Ultimate Causation". In *Georges Lemaître: Life, science and legacy*, Ed. R. D. Holder and S. Mitton (Springer-Verlag GmbH), 125-144.

G. F. R. Ellis, D. Noble, and T. O'Connor (2012) "Top Down Causation: An Integrating Theme Across the Sciences?" *Journ Roy Soc Interface Focus* 2:1-3.

C. Frith (2007), *Making up the mind: How the brain creates our mental world*. (Malden: Blackwell).

A. Gazzaley and A. C. Nobre (2012), "Top-down Modulation: Bridging Selective Attention and Working Memory," *Trends in cognitive sciences* 16: 129-135.

S. Gilbert and D. Epel (2008), *Ecological developmental biology integrating epigenetics, medicine, and evolution* (Sinauer).

J. Kagan (2009), *The three cultures: Natural sciences, social sciences, and the humanities in the 21st century* (Cambridge University Press).

E. Kandel (2012), *The age of insight: The quest to understand the unconscious in art, mind, and brain, from Vienna 1900 to the present* (Random House).

A. McGrath (2011), *Why God won't go away* (London: SPCK)

A. McGrath and J. C. McGrath (2007), *The Dawkins delusion? Atheist fundamentalism and the denial of the divine* (London: SPCK).

N. C. Murphy and G. F. R. Ellis (1995), *On the moral nature of the universe: Theology, cosmology, and ethics* (Fortress Press).

D. Noble (2008), *The music of life: Biology beyond genes* (Oxford University Press) http://musicoflife.co.uk/.

M. Pigliucci (2013), "Ethical Questions Science Can't Answer" http://rationallyspeaking.blogspot.co.uk/2013/10/ethical-questions-science-cant-answer.html

Massimo Pigliucci (2013), "New Atheism and the Scientistic Turn in the Atheism Movement" *Midwest Studies In Philosophy*, XXXVII [http://philpapers.org/archive/PIGNAA.pdf]

D. Purves (2010), *Brains: How they seem to work*. (Upper Saddle River, NJ: FT Press Science).

M. J. Rees (1999), *Just six numbers: The deep forces that shape the universe* (Weidenfeld and Nicholson).

P. Russell (2013), "Hume on Religion." In *The Stanford encyclopedia of philosophy*, Ed. N. Zalta, http://plato.stanford.edu/archives/fall2013/entries/hume-religion.

S. Satel and S. O. Lilienfeld (2013), *Brainwashed: The seductive appeal of mindless neuroscience* (Basic Books).

M. Wertheim (2013), *Physics on the Fringe: Smoke rings, circlons, and alternative theories of everything* (Walker & Co).

8

Owen Flanagan

Owen Flanagan is a world-renowned American philosopher, specializing in philosophy of mind, cognitive science, moral psychology, and ethics. He is currently James B. Duke Professor of Philosophy, Professor of Psychology and Neuroscience, and Professor of Neurobiology at Duke University. His books include *The Science of Mind* (1984, 2nd ed. 1991), *Varieties of Moral Personality: Ethics and Psychological Realism* (1991), *Consciousness Reconsidered* (1992), *Self Expressions: Mind, Morals, and the Meaning of Life* (1996), *Dreaming Souls: Sleep, Dreams, and the Evolution of the Conscious Mind* (1999), *The Problem of the Soul: Two Visions of Mind and How to Reconcile Them* (2002), *The Really Hard Problem: Meaning in a Material World* (2007), *The Bodhisattva's Brain: Buddhism Naturalized* (2011), and *Moral Sprouts and Natural Teleologies: 21st c. Moral Psychology Meets Classical Chinese Philosophy* (2014). Flanagan was awarded a Fulbright Research Award in 2001-2002 to study Buddhist and Hindu conceptions of the self, and in the fall of 2013 he was Distinguished Research Professor at City University Hong Kong and lectured widely in East Asia on 21st century moral psychology and East Asian philosophy.

THE PROBLEM OF THE SOUL

1. What initially drew you to theorizing about science and religion?

Word and Object. The world to which I was introduced as a child included God, angels, saints, and heaven and hell alongside Mom, Dad, grandparents, aunts, uncles, brothers, sisters, houses, furniture, cats, dogs, hamsters, turtles, frogs, and snakes. All the occupants of this great chain of being were equally real and I knew the words for them all right from the get go. It was a perfect laboratory to cause wondering and worrying about what heaven and hell were like, about the considerable differences between parental and divine justice, and about the entrances and exits of souls to and from bodies. Sacred Heart church and grammar school in Hartsdale New York was near Valhalla where all the humans were buried, religion-by-religion (Catholics in "Gate of Heaven"; Jews in "Sharon Gardens"; etc.), and it was directly across the street from America's oldest dog cemetery, "The Peaceable Kingdom" founded in 1896. My parents and the nuns explained that the dogs buried in the dog cemetery being animals "just died" and were in "The Peaceable Kingdom" forever, or just gone or some such, whereas the people buried in "Gate of Heaven" were no longer there, but in heaven.

I remember sometimes envying the dogs, thinking it consoling to die and just be dead after a noble Lassie-like life, as opposed to worrying about the judgment of God and spending eternity figuring out stuff to do in the hereafter. But then again I was worried about the world going on without my knowing how things turned out for those who remained. So eternal life had its appeal.

I seemed to survive more or less psychologically intact puzzling about all sorts of theological distinctions: animals versus humans, venial (everyday selfishness, lies) and mortal sins (sex, always sex), the Immaculate Conception (Mary being born without original sin) versus the Virgin Birth (Jesus being born to Mary without Mary ever having sex), the Trinity, and especially transubstantiation (it seemed extremely silly to believe that Jesus would really come to be in thin wafers and wine). But I recognized even as a boy that wiser souls than me had determined that nothing less that the meaning and significance of everything depended on the complex metaphysics and ethics contained in the form of life we called "being Catholic." This was immensely fertile ground for my philosophical interests in mind, morals, and the meaning of life to take root.

2. Do you think science and religion are compatible when it comes to understanding cosmology (the origin of the universe), biology (the origin of life and of the human species), ethics, and/or the human mind (minds, brains, souls, and free will)?

Abrahamic Myopia. To the best of my knowledge, there is no settled and complete scientific theory about the origin of this universe, the origin of life in it, and then the origin of consciousness in some of the living systems (animals but probably not plants). But there are several good explanation sketches about how these things might have come to be by natural processes. Big Bang cosmology, evolution by natural selection, and naturalism about consciousness are examples. Questions of compatibility need to be taken with respect to particular religions at a particular time. Are we talking about Jews, Christians, Muslims or Buddhists, Jains, Taoists, Hindus or Ecuadorean or New Guinean aboriginals? Traditionally, the Abrahamic religions favor origin stories that involve divine creation of this world and its occupants ex nihilo. But theologians inside these traditions are smart and creative enough to rethink God and his role when these stories compete with science. Perhaps God created this world by way of the singularity that banged 14 billion years ago, which then yielded itself, life, and eventually conscious life, according to the laws of physics, inorganic chemistry, organic chemistry, basic biology, evolution, neuroscience, etc., where the laws themselves emerge in time according to God's plan. This is compatible with science as far as I can tell. Is it plausible all things considered? No, but the reason is that no particular story to fill the holes of our ignorance, or of the mystery, to put it more positively, is plausible given all the possibilities left once science, as developed thus far, has spoken. But that sort of complaint or demand may overstate the degree to which religious stories are intended to provide a true or plausible picture of how things actually happened. Perhaps they are better interpreted as providing a certain sort of possible and uplifting picture that binds a people. The word 'religion' comes from the Latin 'ligare' = to connect or to bind; 're-ligare' = to keep connected. Origin stories that bind peoples need not be understood as place-holders for science; they might be better interpreted as identity-conferring and identity-constituting ways of making sense of things, providing a moral order, and marking and creating solidarity. And, for all I know, the unlikelihood of a story inside the realm of the possible might be an advantage for creating solidarity. It might endow a group with a sense of their own special purpose or mission. Whether this is morally or politically good, indeed whether identification and solidarity are good depends on whether the group has good values, is not xenophobic, etc. But that is a different matter.

Most religions in the world now and in the course of human history, of course, are not the Abrahamic ones even if now on earth approximately half the people identify with one of the Abrahamic religions. Take Buddhism. Traditional Buddhism is atheistic by Abrahamic standards since there is officially no creator God (although some Buddhists believe in one). Most Buddhists are not impressed by first cause arguments, preferring the fork that takes one to a universe that has always existed, an infinite or eternal regress. Some say this makes Buddhist cosmology more compatible with scientific cosmology than the Abrahamic personal creator God picture. Maybe. But traditional Buddhism is chock-full of ghosts, spirits, reincarnated bodhisattvas, and karmic recycling of conscious minds (which are however "no-selves"), according to impersonal laws of karma. So Buddhism is hardly naturalistic despite being atheistic. Perhaps it scores well on the fit of its cosmology with modern science; but its karmic and dualist ontologies are not good fits with contemporary biology, psychology, and neuroscience. Can Buddhism be naturalized and be made consistent with the sciences? Maybe. I suggest a way in *The Bodhisattva's Brain: Buddhism Naturalized*. Does it matter whether the claims of any particular religion can be naturalized? In the Buddhist case, I think it should matter to secular Westerners who are attracted to it and who are also impressed by science whether it has forms that can be understood naturalistically without decimating the Buddhist philosophy in it. But, in general, the answer to the question of whether we should want to tame and naturalize every religion in order to avoid conflict with science depends on whether the claims of any particular religious tradition that are thought to conflict with science are meant as assertions about the way things are and thus actually do conflict with science. If religious claims, texts, rituals and the like are not intended as assertions, but rather as identity-constituting and identity-conferring expressions of solidarity then there is no conflict. That is, if religious assertions are designed to play other roles than description and explanation, such as motivating a moral order, painting an uplifting, beautiful picture of communal purpose, conferring or endowing a sense or purpose on human life that it doesn't, possibly cannot have, when seen from a purely scientific perspective, then they do not in those cases conflict with science. Determining what kinds of speech acts religious people or, for that matter, scientists are making requires sensitive hermeneutics; it cannot be read off the statements of fanatics or fundamentalists.

Overall, my view is that it is relatively easy for most religions to make peace with cosmology, biology, and mind science. But there is sometimes this rub: soteriology or eschatology. Whereas many focus on conflicts about origin questions, there are sources of conflict or tension about afterwards questions. Many forms of Christianity, Islam, Hinduism and Buddhism posit conscious afterlives for persons. Such beliefs are not inconsistent with science in the sense of expressing deep impossibilities. Personal immortality is possible. But if we accept that we are animals, and that consciousness comes with the body, the inference to the best explanation is that there is no such thing. Once again, many sophisticated thinkers across the religions I know about take talk of immortality or eternal life metaphorically, not literally.

3. Some theorists maintain that science and religion occupy non-overlapping magisteria—i.e., that science and religion each have a legitimate magisterium, or domain of teaching authority, and these two domains do not overlap. Do you agree?

Triple Interpenetration. There is something to the idea of non-overlapping magisteria, but the way Stephen Jay Gould ran his argument for non-overlapping magisteria, NOMA, doesn't work. I gave reasons why in *The Problem of the Soul: Two Visions of Mind and How to Reconcile Them.* Gould says that science deals with facts and theories about why the facts are as they are, and that religion deals with values. So they don't overlap. There are several mistakes here. First, some religions or sects do make assertions that do what Gould says only science does, for example, "Genesis" literalists. Second, Gould conflates religion with ethics. Most ethical theories, including ones that are historically embedded in religions, are compatible with science at the level where they endorse certain virtues, character traits, values, and principles. "Be honest" is an imperative and does not compete with any declarative sentence, e.g., "people often lie." "One ought to tell the truth" expresses a norm and does not compete with any fact nor, for familiar reasons does it deductively follow from facts such as "lives go better for truth tellers." But many, probably most, moral traditions locate the source of moral values in some metaphysically deep supernatural place (Plato discusses this commonplace in the "Euthyphro"), which for reasons given above need not involve a creator or a personal God. It—the deep source—can involve a karmic system where the universe itself knows and keeps track of right and wrong, good and bad as in Hinduism, Jainism, and Buddhism; or it can involve a natural teleology, a certain directionality in nature or heaven, as one sees in certain Chinese spiritual traditions such as Confucianism and Taoism.

The project of the enlightenment in the North Atlantic is devoted to the attempt to make sense of our values without the supernatural scaffolding that had historically co-constituted the ethics and religion of the Abrahamic traditions. So both Hobbes and Mill explain that the ethics they endorse is well summed up by the Golden Rule of Jesus of Nazareth. This is very different than saying that their ethics are justified by the fact that Jesus of Nazareth gave them pithy expression. It is an interesting and important philosophical question, as well as an interesting and important sociological and political one whether ethics without supernatural sources can be philosophically or sociologically robust. It is a big mistake to conflate the two questions. I think we know the answer to the first question: We can often make sense of and justify our values naturalistically. I do not think we know the answer to the second question, whether there can be social and political order (and if there can be, when, how soon, under what social, political and economic conditions), without some sort of belief in supernatural sources, or what is different, without solidarity around stories that express in spiritually and aesthetically uplifting forms the multifarious commitments of a people.

4. What do you consider to be your own most important contribution(s) to theorizing about science and religion?

Non-Abrahamic Traditions & the Functions of Religion. I recommend two ideas. First, religion is not one thing. Religions are multifarious in kind. Many have non-Abrahamic structure and character. I've been writing a fair amount recently about Buddhism (all experts agree that Buddhism despite being atheistic is a religion) and Confucianism (some say Confucianism is a religion; some not). They are deeply different from each other and from the religions of Abraham. Some religions have no texts dictated by God, some have no texts at all. Some religions don't have a personal, or, what is different, a creator God. Many religions have gods who are finite and flawed, not omni-this and omni-that. Many religions are not creedal. Some religions are highly ritualistic and involve altered states of consciousness. This is only the beginning. But acknowledging this much will make the philosophy of religion less parochial and more psychologically, sociologically, and anthropologically realistic. It will also make it more obvious that the conflict between science and religion is not one single, clear-cut conflict. 'Science' and 'religion' are very general names for complex, fluid institutions, ways of thinking, speaking, and importantly marking communal relations and membership. They do not speak for themselves. Humans speak in, or claim to speak in, the voice of science or religion, sometimes both at once. Whether smart or foolish things are being said in the name of either needs to be determined on a case-by-case basis, or perhaps, if one is a pragmatist, by the net positive or negative effects of the two very general forms of inquiry, speaking, and being in specific cultures at specific times. I seriously doubt that there is much that is both very general and very helpful to be said about what science and religion are and about their relations.

The second idea that I think is helpful is related. In *The Really Hard Problem: Meaning in the Material World* (2007), I recommend a distinction between "assertive theism" and "expressive theism." The idea is that it is no easy task to judge whether to interpret some set of religious claims as assertions or as some other kind of speech acts. What other kinds? There are many possibilities. Origin narratives of the universe or a people can be read as attempts to arouse and then center the heart and mind, to think and feel awe or beauty or solidarity, or all these at once. Stories of saints, bodhisattvas, and sages can be read as providing moral ideals, high aspirations, or visions of excellence for a people. Shamans, gurus, priests, rabbis, and prophets make recommendations for self-cultivation, for cleansing oneself of destructive emotions and for political action. Sometimes they are full of shit, scary and dangerous; other times they speak truth to power. There are meditation techniques designed to feel one's way out of the narrow selfish ego; there is prayer and chanting, and there is LSD, ayahuasca, and peyote to make one feel more at one with the universe. Is there a mistake, a factual mistake, a scientific mistake in feeling at one with the universe? What if one actually believes that one is connected to the universe. Is this an illusion or delusion? I doubt it. But it raises interesting questions about the nature and contours of the self, who speaks most truthfully about it, and whether there is one and only one way to understand the self. I leave it for homework.

5. What are the most important open questions, problems, or challenges confronting the relationship between science and religion, and what are the prospects for progress?

Enchantment, Disenchantment, and Platonic Unification. If I had one recommendation to make to some fellow philosophical naturalists who work this terrain it would be to show more humility about the resources of science to provide sources of value and grounds for meaning and purpose. At its extreme this view is scientism, the view that all truths worth expressing are best expressed in a scientific idiom. Most people on earth are religious. Some religions at some times endorse false or irrational beliefs. Sometimes these are dangerous; other times they are harmless. But, to take one prominent example, the belief that God might have created the world to work by way of evolution is not one of the obviously false or irrational ones. And calling people stupid or ideologues that believe it shows neither humility, nor compassion, nor civility, nor a sense of the ongoing difficulty of explaining our place in the universe. Religions have been good historically at providing comprehensive ways of conceiving of a world, of locating meaning and purpose, of making an ethical form of life explicit, and of finding and creating a community for celebration and consolation. One big open question is whether a fully secular naturalistic picture of reality can be satisfying. Satisfaction can be understood in terms of truth of the sort that the sciences specialize in or it can be understood more widely in terms of what I call "platonic unification." Plato had it right that a good life is lived at the intersection of what is true, what is good, and what is beautiful. It is obvious by this time in human history that it is not easy to make these goods perfectly harmonized. But it is a worthwhile project. It may just be that calls for a fully scientific view of reality overly prizes the true, and indeed the truth of a certain kind, and underestimates the need for goodness and beauty as well. Music does not speak in a scientific mode; but many of us thinks it expresses certain kinds of truth and that human life would be impoverished without it. I am inclined to think that the religious impulse is to be respected, understood, and appreciated as a complex way of thinking, speaking, and expressing the human commitment to do what Plato said was our nature, seeking maximal intersection of truth, beauty, and goodness in our individual and collective lives. Understanding better this meaning-making role, these meaning-making roles, of religion and science, seems to me the next right thing; better at any rate than intemperate and disrespectful trash-talking under the cloak of science or "the truth." Truth isn't, I dare say, the only thing that matters, or perhaps more judiciously, if you already think that music, art, and ethics speak powerfully and truthfully about reality then you already think that there is more worth saying, thinking, and being than science alone offers.

9

Owen Gingerich

Owen Gingerich is Professor Emeritus of Astronomy and of the History of Science at Harvard University and a senior astronomer emeritus at the Smithsonian Astrophysical Observatory. He is the co-author of two successive standard models for the solar atmosphere and is a leading authority on the 17th-century German astronomer Johannes Kepler and the 16th-century cosmologist Nicholas Copernicus. In recognition of his work, he was awarded the Polish government's Order of Merit in 1981, and subsequently an asteroid was named in his honor. Professor Gingerich has been Vice President of the American Philosophical Society (America's oldest scientific academy), has served as Chairman of the US National Committee of the International Astronomical Union, and has been a councillor of the American Astronomical Society (AAS). In 2004 the AAS awarded him their Education Prize, and in 2006 he received the most prestigious award of the French Astronomical Society, their Prix Janssen. Besides two hundred technical or research articles and three hundred reviews, Gingerich has written more than two hundred and fifty education, encyclopedia, or popular articles. He is also the author of several books, including *The Eye of Heaven: Ptolemy, Copernicus, Kepler* (1993), *An Annotated Census of Copernicus' De revolutionibus (Nuremberg, 1543 and Basel, 1566)* (2002), *The Book Nobody Read: Chasing the Revolutions of Nicolaus Copernicus* (2004), and *God's Universe* (2006).

---

1. What initially drew you to theorizing about science and religion?

Every thoughtful person faces a challenging existential question that requires a decision—whether consciously or unconsciously: Is the universe purposeful? Or is it meaningless, full of sound and fury, signifying nothing? As I have often said, I'm psychologically incapable of believing that the universe is meaningless, and therefore I accept that the universe is purposeful even if I can't clearly define what the purpose is. But I suppose it must have something to do with the thinking beings that inhabit it, their relations to each other, and to the larger universe in which they dwell. The mere fact that the universe seems to be understandable suggests that a supremely intelligent creator is behind it all, and at least part of the purpose lies in the challenge to decipher the mystery of the universe.

My beliefs were rooted in a religious upbringing in small-town Midwest. My father, with a PhD in American history, was a voracious reader who must have over the years reviewed well over a thousand books for a column in the *Mennonite Weekly Review*. His wide interests included a serious dose of theology. My undergraduate college had the motto

"Culture for Service," and I had to wrestle with the idea of graduate work in astronomy, which seemed rather removed from "service," but my mentors said we shouldn't let the atheists completely take over any area of knowledge.

I was always intensely interested in the nature of science itself, the basis for its claims to truth, and the interaction between hypothesis and observation. Doing science brought me face-to-face with issues of what to pursue and what to ignore in trying to decode the universe. Eventually I realized that a detailed examination of some of the historical record offered a rich perspective on the ways science works.

With this background I suppose it was inevitable that I gradually became an occasional essayist in the interplay between science and religion. This led, for example, to an essay on "Kepler's Anguish and Hawking's Query: Reflections on Natural Theology" in *Great Ideas Today 1992* (the yearbook of *Great Ideas of the Western World*), with philosopher Mortimer Adler as commentator. Around the same time an earlier essay, "Let there be light: Modern cosmogony and Biblical creation," was printed in a Book of the Month Club selection, Timothy Ferris' *World Treasury of Physics, Astronomy, and Mathematics*.

2. Do you think science and religion are compatible when it comes to understanding cosmology (the origin of the universe), biology (the origin of life and of the human species), ethics, and/or the human mind (minds, brains, souls, and free will)?

The scientific picture of the universe established over the past two centuries depicts an unimaginably ancient structure with a history. Contrary to Aristotle, the universe we observe had a beginning and has not existed forever. It was an immensely long and gradual process of forming planets and the atoms required for life. In contrast, the Biblical account is of a sudden creation of the world, not all that long ago. It seems to me there are two ways to reconcile these very diverse creation stories. One is to suggest that the world is only 6,000 years old, but all these ancient features like fossil dinosaurs or emergent galaxies seemingly 12 billion light-years away were simply built into the universe at that moment to challenge us and to test our faith. That would be a trickster god, a concept that challenges my credulity.

The alternative is to seek a non-scientific purpose for the first chapters of Genesis, namely, to distinguish the Hebrew tradition of a monotheistic Creator from the rival polytheistic creation myths of the surrounding cultures. The Genesis account is the story of the emergence of human conscience and responsibility, a message that resonates all the more expressly today when we hold the power to destroy not only our environment but the entire human race.

Are religion and science compatible in our age of science? Clearly *no!* if we attempt to read scripture as literal history, including the story of a universal flood. The authors and editors of the scriptures could well have been inspired in godly intents even if they were not writing history. The answer is *yes!* if we accept that the scriptures convey truth just as great music, art, or even fiction writing can do without recounting actual historical circumstances.

3. Some theorists maintain that science and religion occupy non-overlapping magisteria—i.e., that science and religion each have a legitimate magisterium, or domain of teaching authority, and these two domains do not overlap. Do you agree?

The expression "non-overlapping magisteria" arose from my late friend and colleague, the paleontologist Stephen Jay Gould, in his book *Rocks of Ages*. In principle, the idea that science and religion operate in two different domains and they will get along quite compatibly if each sticks to its own magisterium seems like a good idea. But if you look at this historically, or even in the present-day structures, you will quickly find that there has always been, and continues to be, a significant overlap.

Recently I analyzed this situation through a series of three case studies entitled "Was Copernicus right?" "Was Darwin right?" and "Was Hoyle right?" and these are being published by Harvard University Press as *God's Planet*. Let me mention two contemporary situations from the third case study where magisteria overlap.

The first is the so-called fine tuning of the universe. Our universe of planets, stars, and galaxies depends on certain physical constants having values within a limited range close to what we actually find. In particular, the resonance levels in the carbon and oxygen nuclei predicted by Fred Hoyle have to be well tuned to make carbon plentiful enough for the structures of life. As Hoyle said, either there have to be monstrous coincidences or else a superintelligence designing the universe.

As a theist who believes in a purposeful universe, I accept as a final cause a Creator-God being a designing superintelligence. This does not answer to the efficient cause, beloved by modern science, that explains *how* it was done. An increasingly popular idea is that our universe is simply one of many with assorted physical constants, and naturally we would have to be in one with the physical constants set so that intelligent life could arise therein. In other words, our universe is one that came out randomly in the grand cosmic roulette, hence accidentally rather than designedly. Of course this still does not explain *how* it was done and what guaranteed it to have at least one among the trillions that would be favorable to life. Accepting this as the scheme to explain that ours is not a purposively designed universe means accepting another non-scientific principle, arising from an overlapping magisterium. Why is it non-scientific? Because these multiple universes are either unobservable because they are in their own independent non-connecting spaces, or else they are in unobserved portions of our own universe, with entirely different laws of physics. So far there is not a shred of observational evidence that multiverses exist, despite the popular literature. Faith in the existence of other universes is surely coming from an overlapping magisterium of fundamental beliefs, just as is the belief that the universe has been specifically designed to support life complex enough to think religious or philosophical thoughts.

There is a parallel situation with respect to possible life elsewhere in our universe. When Copernicus argued that the earth is one of several planets rather than being the unique central object in the universe, he opened the door to speculation that other planets might also be inhabited. If the conditions are right, inhabitants would surely be present. This is the principle of plenitude, already discussed by Plato, but hardly a rule situated in the magisterium of science. It is a philosophical opinion, a mirror of our own view of ourselves, which might someday prove to be correct, but at present is mere speculation.

The lure of plenitude has brought many distinguished astronomers under its spell, from Kepler and Huygens to Herschel, Percival Lowell, and Carl Sagan. In the 19th century Laplace's nebular hypothesis held sway as the way planetary systems formed, and in this scenario planetary systems could be an abundant result of stars (and planets) forming by the condensation or accretion of massive clouds of gas and dust. Eventually this picture failed because the sun was rotating too slowly to have a proportionate amount of angular momentum, and in its place came a collision theory, in which two stars nearly collided, pulling off gigantic streams of gaseous material from which planets could form. Because the stars of the Milky Way are so widely separated, stellar collisions in our galaxy would be exceedingly rare, and our own planetary system almost unique. Thus in the first half of the 20th century, visions of other inhabited worlds were virtually limited to Mars.

In the 1950s, however, the nebular theory was dusted off, and the sun's slow rotation was attributed to secondary effects after the initial formation of the planetary system. Consequently, the idea of innumerable planetary systems became almost a given in thinking about the nature of stellar system, something brilliantly confirmed observationally in the past two decades. It will not be a surprise if an oxygen signal, a clue to the existence of extrasolar system life, is detected in the next ten or 15 years. This, however, will not reveal anything about the level of life, and the notion of intelligent life elsewhere will remain an overlapping magisterium from an overarching philosophical view, conceivably with roots in a scientific understanding of what is possible, but nevertheless taking its nourishment from a broader view of what the universe should be like. In this it is more akin to theological views than scientific ones.

4. What do you consider to be your own most important contribution(s) to theorizing about science and religion?

My interests have encompassed both modern cosmology and the historical origins of modern science. I am sympathetic to theological approaches to understanding the place of humanity in the cosmos, but I always consider myself only an amateur theologian. Nevertheless, I have been in a position to bring elements of modern science together with insights into the Christian background and context of pioneers such as Copernicus, Kepler, and Galileo. This in turn has given me a background to speak about evolution, which is hugely controversial and much misunderstood among many Christians today. I am currently writing especially in this domain.

5. What are the most important open questions, problems, or challenges confronting the relationship between science and religion, and what are the prospects for progress?

Science has brought us into the modern world. There is an anecdote, probably mythological, that the English prime minister asked Michael Faraday what use electricity could have, to which he responded, "Someday you will tax it." We can hardly envision a world without electricity. That smallpox has been eliminated, and polio nearly so, are remarkable features of today's world. But so are nuclear weapons, the dangerous possibilities of biological warfare, the way we nearly destroyed the ozone shield (which could be attributed to 20th-century chemistry).

Most religions, I am told, contain something equivalent to the so-called Golden Rule. Jesus said, "So whatever you wish that others would do to you, do also to them." This is the central ethic of Christianity. Yet, as we read the daily *International New York Times*, we see that the world is very far from accepting this rule. The New Testament speaks much about salvation, and if we ask, "what can be the salvation of the human race?," much depends on taking the Golden Rule very seriously indeed.

Clearly both science and religion are very powerful forces in our world today, and not necessarily forces for the survival of humankind. The more closely science and religion can be harnessed to work together, the better our prospects are. But what the prospects for progress are, I cannot say nor, alas, can I be very optimistic.

10

Rebecca Newberger Goldstein

Rebecca Newberger Goldstein is an American philosopher and novelist. She has published seven books of fiction, and three non-fiction books. The titles are: *The Mind-Body Problem* (1983), *The Late-Summer Passion of a Woman of Mind* (1989), *The Dark Sister* (1993), *Strange Attractors: Stories* (1993), *Mazel* (1995), *Properties of Light: A Novel of Love, Betrayal, and Quantum Physics* (2000), *Incompleteness: The Proof and Paradox of Kurt Gödel* (2005), *Betraying Spinoza: The Renegade Jew Who Gave Us Modernity* (2006), *Thirty-Six Arguments for the Existence of God: A Work of Fiction* (2010), and *Plato at the Googleplex: Why Philosophy Won't Go Away* (2014). Her Tanner Lectures on Human Values, delivered at Yale, were published as *The Ancient Quarrel: Philosophy and Literature* (2012). Her books have won numerous awards, including the Whiting Writer's Award, the National Jewish Book Award (1995), the Edward Lewis Wallant Award (1995), and the Koret International Jewish Book Award in Jewish Thought (2006). She has received fellowships from the Guggenheim Foundation, The American Council of Learned Societies, the Radcliffe Institute, and the Santa Fe Institute. In 1996 Goldstein became a MacArthur Fellow, receiving the prize popularly known as the "Genius Award." In 2005 she was elected to The American Academy of Arts and Sciences, and in 2008 she was designated a Humanist Laureate by the International Academy of Humanism. In 2011 she was named Humanist of the Year by the American Humanist Association and Freethought Heroine by the Freedom from Religion Foundation.

---

1) Why were you initially drawn to intellectual history?

My interest in science derived much of its passion from religion. The passion was a reaction against religion. I was born into a religious family, and the claims of family and teachers—about the cosmos, morality, human history and destiny—left me with one over-riding question: how can these claims be known, much less known with such certainty? They didn't seem at all obvious to me. What wasn't I getting? I asked a lot of questions, and they tended to be of the *how-do-you-know-all-that-is-true* variety. I was often assured that people smarter than I had answered any question I could come up with, which response was offered in lieu of more satisfying answers.

I spent a lot of time at the public library. This was where I discovered science books for children, including *Our Friend the Atom*, by Heinz Haber. *Our Friend the Atom* blew my mind. I couldn't get over the fact that the solid, sluggish world was made up of all these frantically moving bits. Once again was the question: *how do they know all that is true*? I formed an ambition, appropriately modest. Someday I was going to get to meet a scientist. I wanted someone to explain to me not only how the invisible world of the atom was known but also how that invisible world could result in the world that I saw and felt. Was the world I saw and felt all wrong? Were colors not real, since our friend the atom was colorless? But colors seemed so real! Those were the questions I was going to ask if I ever got to meet a real live scientist.

I liked the kind of explanations I got from children's science books, and, one by one, I checked them all out from our local library. One of my favorites was entitled—or so I remember—*Turn Your Mother's Kitchen into a Science Lab*. It came with explanations for why heating up sugar eventually turned the white powdery stuff into a glob of black stickiness, or why, if you inserted a burning piece of paper in the bottom of a glass milk bottle with a narrow neck on which you quickly perched a hard-boiled egg, the egg would get forced into the bottle. I loved these explanations because they let you see through the visible to such invisible things as a partial vacuum and air pressure. Understandably though, my mother wasn't thrilled about what happened in her kitchen, and, not even halfway through it, I wasn't allowed to renew that particular book.

Knowledge is a hard thing to come by. That was the fact that the children science books impressed on me. You have to keep asking, and you have to keep testing. Even when very smart people—scientists!—think they've got something right, they often don't, *and then they change their minds*. Nobody I knew seemed interested in a mind-change. I wanted to be the kind of person who could change her mind.

Eventually, when I got to college, first physics and then philosophy of science attracted me. I was particularly interested in the question of how to reconcile experience—including the fact of experience itself—with the scientific picture. I was drawn to Wilfrid Sellars's characterization of philosophy (in "Philosophy and the Scientific Image of Man") as providing the coherence-expanding mediation between the *manifest* and the *scientific* images of what he called "man-in-the-world." His conception of philosophy has remained vital to me.

The character of my family has kept me intrigued by the nature of religious belief, as well as disinclined to reject believers as intellectually deficient, as they are too often so characterized by those who share my lack of religious belief. My religious family is anything but intellectually deficient. They are highly critical thinkers, as skilled in the art of ingenious disputation as any analytic philosopher I've encountered. But there is a domain of beliefs to which their honed critical skills are never applied. It's not even that they don't argue for these beliefs; it's that they're not prepared to argue for them—*deliberately not prepared*, they who are prepared to argue any point. Their deliberate unpreparedness speaks to the fact that their religiosity, as far as I have come to understand it, is not a matter of theory but rather a way of life, in a way that reminds me of certain statements of Wittgenstein's: "It strikes me that a religious belief could only be something like a passionate commitment to a system of reference. Hence, although it's *belief*, it's really a way of living, a way of assessing life. It's passionately seizing hold of this interpretation" (*Culture and Value*, 64c). Because their religiosity doesn't function at the level of theory, they see no incompatibility between science and religion; there is no scientific theory they would reject on religious grounds. There are, no doubt, beliefs underlying their way of life—perhaps even beliefs about God, though concerning these, they are reticent. Of far more importance to them is the character of the life that results from the intensity of religious rituals performed. Out of the performance emerges a life that, for them, justifies itself. From their point of view, my *how-do-you-know-all-that's-true* variety of questions naively misses the point. From my point of view, their interrogatory insouciance is epistemologically irresponsible. It's this kind of disagreement, concerning the ethics of knowledge, that interests me more than tensions surrounding science and religion that take place on the theoretical plane.

2. Do you think science and religion are compatible when it comes to understanding cosmology (the origin of the universe), biology (the origin of life and of the human species), ethics, and/or the human mind (minds, brains, souls, and free will)?

I think that religion, when it goes beyond a way of life and ventures into theory to make claims about the nature of the world, is fundamentally incompatible with science, no matter the claims made. The incompatibility exists at the level of epistemology. Sometimes the fundamental epistemological incompatibility is compounded by specific instances of factual incompatibility, as when, say, miracles are affirmed, or the geological age of the earth or the evolutionary origin of the human species is denied. Almost all religions have some version of a theory of an immaterial soul functioning as a radically free moral agent to be held accountable in an afterlife. All of these aspects of the soul theory—immateriality, radical freedom, afterlife—seem to me to be in dissonance, if not strict irreconcilability, with neuroscience. But even when factual religious claims about the nature of the world are sufficiently tamed so as to be as reconcilable as possible with science, the two, religion and science, are animated by irreconcilable epistemologies, which is why I say religion, if understood as offering claims about the world (onto-religion), is fundamentally incompatible with science.

Science takes seriously the fact that our brains are not the most efficient cognitive mechanisms. Advances in cognitive science and evolutionary psychology are only now beginning to explain why our intuitions and our reasoning so often lead us astray. But even before we had any scientific explanations for our systematic fallibility, modern science was acknowledging that fallibility in the stringency of its methodology. The great trick of the empirical sciences is to give nature a chance to answer us back when we get it hopelessly wrong. Oh, so you think that simultaneity is necessarily absolute, do you, no matter what inertial frame of reference it's measured in? Well, we'll just see about that! Voila, the Special Theory of Relativity.

What counts as justification for beliefs within onto-religion obviously doesn't meet the demanding criteria of science; if it did, it would be science. Onto-religion doesn't meet science's criteria for how best to defeat our impressive powers for getting ourselves fooled about the nature of the world, and it doesn't offer any comparably stringent criteria of its own to impede our fallibility. In fact, some of what counts as justification in religious contexts—the authority of ancient texts and of religious authorities, for example, not to speak of faith—significantly up the chances for getting everything about the world wrong.

There are, of course, questions that science still hasn't answered, and some are of so fundamental a nature as to suggest that they might exceed, in principle, the methods of science. I have in mind such a question as: Why does the universe have the specific laws of nature that it has? We explain phenomena by subsuming them under laws of nature and explain laws of nature by subsuming them under more fundamental laws of nature, but what methods could science use to explain why those fundamental laws of nature are the ones realized in the universe? The theory of the multiverse presents itself as a possible way of providing an answer: There are a vast multiplicity of universes, characterized by different laws of nature, and we just happen to find ourselves in the universe characterized by the laws of nature that supported the evolution of creatures like ourselves, capable of posing this question about the laws of nature that we discover. It still seems to me that the theory of the multiverse presumes fundamental laws of nature, shifting their applicability to the multiverse itself, and so the theory doesn't rid us of the fundamental question. But even if it does, could the theory of the multiverse provide the answer to yet another fundamental question: Why is there a universe—whether multiverse or not—at all? Or, as Einstein put the question: Why did the universe go to the bother of existing in the first place? It's not inconceivable that science will eventually answer even this last question, perhaps by showing us why the most fundamental laws of nature, whether of the multiverse or not, are of such a kind that they *demand* realization. But, in any case, we're not yet in a position to scientifically answer this question and maybe never will be.

Onto-religion rushes in to theorize in the theoretical vacuum our present science still leaves. I believe it's always a mistake for religion to theorize. Religion has given us no reason to accept the reliability of its claims about the nature of the world, certainly nothing on a par with science, with its methodological acknowledgment of our systematic fallibility. Nor for that matter has religion given us reason to accept the reliability of its grasp of ethical truths, certainly nothing on a par with moral philosophy, which also acknowledges our systematic fallibility and has devised its own methods in the light of this acknowledgement.

In the orthodoxy in which I was raised it was accepted as a truism that the further we get from hallowed antiquity the more corrupted our knowledge becomes. This is the very opposite of the history of knowledge that science and philosophy tells.

3. Some theorists maintain that science and religion occupy non-overlapping magisteria—i.e., that science and religion each have a legitimate magisterium, or domain of teaching authority, and these two domains do not overlap. Do you agree?

Those who maintain the non-overlapping magisteria, or NOMA, usually cede to science the magisterium of facts, and they cede to religion the magisterium of values. This was how Stephen Jay Gould, who introduced NOMA, divvied up the territory. I don't mean to sound harsh, but it seems to me that only someone ignorant of philosophy—and in particular, of moral philosophy—could so decree. The territorial division implies that there are no facts of the matter about values, no facts of the matter about what—and who—matters. You might as well let the religious authorities decide such questions, since there's really nothing to know, no possible error to be made. If the Church decides to burn heretics, based on the authority of John 15:6—"If a man abide not in me, he is cast forth as a branch, and is withered; and men gather them, and cast them into the fire, and they are burned,"—there's no fact of the matter that this is wrong, so let the Inquisition commence. Or if, on the basis of *Leviticus* 25:44-46—which authorizes treating non-Israelite slaves as heritable property—or of *Ephesians* 6:5—exhorting slaves to obey their masters with the sincerity of heart due to Christ himself— where could the fallacy lurk? There's no fact of the matter regarding whether some humans are born with a right to own their own lives while others aren't.

Only there *are* moral facts of the matter, and so there are mistakes to be made—mistakes that have robbed untold numbers of people of lives worth living, and it has taken the rigorous arguments of philosophical reasoning to demonstrate the indefensibility of practices protected by millennia of tradition. Here, for example, is Sebastian Castellio, in

1553, arguing against the burning of heretics: "Calvin says that he is certain, and [other sects] say that they are. ... Who shall be judge? ... If the matter is certain, to whom is it so? To Calvin? But then why does he write so many books about manifest truth? ... In view of the uncertainty we must define the heretic simply as one with whom we disagree. And if then we are going to kill heretics, the logical outcome will be a war of extermination, since each is sure of himself." Castellio was a preacher and theologian but the argument he is making is not theological but philosophical. Likewise, the arguments against slavery originated in secular philosophical arguments, with pride of place for the first anti-slavery argument usually going to the Enlightenment's Montesquieu, who in Book 15 of *The Defense of the Spirit of the Laws* (1750) argued the evil of slavery. However, Jean Bodin, the French humanist and jurist, actually beat Montesquieu by one hundred and eighty years, forcefully arguing the immorality and irrationality of the practice in Book One of *Six Books of The Commonwealth* (1576). So it has been with perhaps every advance in human rights: it took rigorous argument to pierce through the daze of morally-blinding tradition and demonstrate that certain attitudes and practices were inconsistent with principles held to be inviolable *when applied to oneself and one's own kind*, and that so limiting the applicability of these principles made for moral incoherence.

Reason sells itself short when it forgets it has resources beyond those of the empirical sciences. Philosophy's mandate is to maximize the coherence of our lives, including our moral coherence. The doctrine of non-overlapping magisteria, in arrogating to the empirical sciences all discoveries, not only reveals itself to be blind to a different order of discovery, made in the service of coherence, but also fails to see the role that reason plays in forcing such discoveries out into the open. The demonstrativeness of philosophical arguments eventually reworks our moral "intuitions," so that, with time, people forget that forceful arguments had once been required to see the intolerability of such practices as burning heretics and keeping slaves. Eventually even religious authorities end up liberalizing their interpretations of passages like *John* 15:6, *Leviticus* 25:44-46, and *Ephesians* 6:5. Perhaps someday the liberal interpretation of such anti-gay passages as *Romans* 1:26 will be similarly de rigueur.

4. What do you consider to be your own most important contribution(s) to theorizing about science and religion?

I suspect that part of what made William James so astute a psychologist of religion was that he always wanted to be a believer but couldn't quite manage the leap. It was James, of course, who coined the term "leap of faith." He also coined another insightful term, "the will to believe," and for most of my life I've thought about religion in terms of the will to believe, connecting it with something I called "the ontological urge," the human desire to figure out what kind of place our universe is. Where, exactly, *are* we? Like many secularists, I've looked to science and sound epistemology to tame the wild excesses to which the ontological urge and the will to believe can lead. The universe doesn't make itself easy for us to know it, and the ontological urge should look to science to satisfy itself.

But recently I've been thinking about religion in relation to an altogether other human will, one which characterizes all of us, whether religious or secularist, and that's the will to matter. I think the will to matter has more to tell us about the persistence of religious frameworks that emerged so many millennia ago and which—unsurprisingly, considering their antiquity—get so much wrong. These theologico-normative frameworks emerged in response to the profound need that each of us feels to matter. Significantly, moral philosophy itself emerged out of the same will to matter. I explore these ideas in *Plato at the Googleplex: Why Philosophy Won't Go Away*.

The religious frameworks that still flourish all emerged in the same period, 800 to 200 B.C.E. The philosopher Karl Jaspers dubbed this the "Axial Age," because its ideas continue to radiate out into our own day, still commanding the assent of millions. This was the period that also saw the emergence in the Greek world of both secular philosophy and dramatic tragedy. Confucius, the Buddha, Ezekiel, Pythagoras, and Aeschylus were all contemporaries of one another. I believe that the intense normative theorizing of this period was a response to political and economic changes, all of them connected with the emergence of large polities in all the regions involved in the normative ferment of the Axial Age. These changes worked to force a preoccupation with individual mattering into consciousness. The Greeks were quite distinctive in their secular approach to the question of human mattering (even though their culture was saturated with religious rituals). Even pre-philosophically, as far back as the Homeric Age, they thought of mattering as something that had to be achieved by doing something so glorious, as to make one's name repeated on many lips, an acoustic renown (or *kleos*) that would ideally outlive one's own brief life. Their approaching the problem of mattering in human terms developed into what I call an Ethos of the Extraordinary. It's only the ordinary life that's not worth living— a harsh statement, but the Greeks hardly recoiled from harshness. When Plato has Socrates declare in the *Apology* that the unexamined life is not worth living, he is addressing the preoccupation with mattering that underlies all the traditions forged in the Axial Age, and he is also being very Greek in conceiving of mattering as something that has to be achieved on human terms. But it isn't our name on many lips that constitutes our mattering but rather our exertions in reason. A harsh Greek presumption thereby slipped unexamined into Plato's reasoning. Some are still inclined to piously regard Socrates' declaration regarding the unexamined life, but, like so many other pieties, this one, when examined further, reveals its grotesqueness.

But the more important point is that, by proposing such an answer to the Axial Age's preoccupation with mattering, Plato initiated the only one of the normative frameworks of that age that had the capacity for self-correction, just as science does. Eventually, moral philosophy self-corrected to the point of eliminating, at least in theory, the unexamined Greek presumption that had kicked off the whole process of applying reason to normative questions. All modern moral theories, whether utilitarian or not, presume an equality of mattering, though in practice the elimination of differential mattering is still a work in progress, and this means that far too many people still have lives that are treated as mattering much less than the lives of others.

Secular moral philosophy can't offer the kind of cosmic mattering that the Abrahamic religions offer, a god keeping tabs on every one of our actions so that we can be called to account at the end of our days—a terrifying proposition, yes, but one that doesn't let a believer doubt for a moment how profoundly she matters. But what secular moral philosophy *can* offer are sound arguments for convincing us that if any one of us matters—and just try to pursue your life coherently without presuming that you do—then all of us matter. I think that mattering is a potent concept, one that bridges the so-called gap between 'is' and 'ought' and gives the lie to NOMA.

5. What are the most important open questions, problems, or challenges confronting the relationship between science and religion, and what are the prospects for progress?

When religion fashions itself as theory, aiming to fill in the answers still missing from science or, far worse, rendering its own "corrections" to answers already supplied by science, it so exceeds its proper boundaries as to not even manage to be wrong.

But there are also questions that might exceed the proper boundaries of science. I referred above to two outstanding questions: Why does the universe have the fundamental laws that it has and why did the universe bother to go through the bother of existing in the first place? For many, the theory-vacancy indicated by these two unanswered questions opens up space for religious answers to move in—a fallacy, by my reckoning, but still it is a move that scientific incompleteness theoretically leaves open. Were science to answer these two questions—in the way, say, that it has answered the question of the origin of the species, with an explanation so satisfactory as to eliminate all scientific rivals—then this would close this theoretical vacancy. What is demanded is a final theory of everything which, explaining everything, explains why it is the final theory, why the laws of nature it describes are sufficiently unique to have forced themselves on the universe and, in so doing, forced the universe into being. Such a theory would allow the universe to render a full accounting of itself by itself. Science would have rendered redundant any proposed supernatural answers to theoretical questions. Onto-religion would then commit its adherents to the kind of anti-science of fundamentalists who now deny the theory of evolution.

Science has gotten far in the progress of knowledge by constantly demanding that the universe accounts for itself. An observed violation of an accepted law of nature is never interpreted as a miracle, but rather as evidence that the law has not been formulated correctly, either because the underlying theory is wrong (as the phlogiston theory of fire is wrong) or is only a limiting case (as Newtonian mechanics is only a limiting case). The progress science has made by way of pursuing the intelligibility of the universe offers some evidence for the belief that the universe is intelligible through-and-through; it can account for itself as completely and conclusively as is demanded by the (awaited for) final theory of everything.

Of course, given our present knowledge, it's possible that the universe is not characterized by such thoroughgoing intelligibility and hence that no final theory of everything awaits our discovery. It's possible that, when it comes to the most fundamental laws of nature and to the existence of the universe characterized by those laws, there is nothing but brute contingency: There are no reasons at all for why there is anything and why it has the structure that it does. But the intuition of cosmic intelligibility that we have been following ever since the awakening of scientific reasoning in ancient Ionia has reaped such an abundance of knowledge, much of it now so counter-intuitive as to tax our powers of imagining, that it gives me reason to believe that intelligibility penetrates all the way down. One can think of cosmic intelligibility as a working hypothesis that has wildly paid off. My belief is that the physical universe is all that we have and that it is entirely enough.

11

John F. Haught

John F. Haught is Distinguished Research Professor at Georgetown University. He specializes in systematic theology, with a particular interest in issues pertaining to science, cosmology, evolution, ecology, and religion. He has authored numerous books and articles, including *The Cosmic Adventure: Science, Religion and the Quest for Purpose* (1984), *What is God?* (1986), *What is Religion?* (1990), *The Promise of Nature: Ecology and Cosmic Purpose* (1993, 2nd ed. 2004), *Science and Religion: From Conflict to Conversation* (1995), *God After Darwin: A Theology of Evolution* (2000, 2nd ed. 2007), *Purpose, Evolution and the Meaning of Life* (2004), *Is Nature Enough? Meaning and Truth in the Age of Science* (2006), *Christianity and Science: Toward a Theology of Nature* (2007), *God and the New Atheism: A Critical Response to Dawkins, Harris, and Hitchens* (2008), *Making Sense of Evolution: Darwin, God, and the Drama of Life* (2010), and *Science and Faith: A New Introduction* (2012). In 2002 Haught received the Owen Garrigan Award in Science and Religion, in 2004 the Sophia Award for Theological Excellence, and in 2008 a "Friend of Darwin Award" from the National Center for Science Education. He also testified for the plaintiffs in the Harrisburg, PA "Intelligent Design Trial" (*Kitzmiller et al. vs. Dover Board of Education*).

---

1. What initially drew you to theorizing about science and religion?

In my early 20s, I started reading the works of the Jesuit priest and geologist Pierre Teilhard de Chardin, and ever since then his vast writings have been a stimulus to my ongoing interest in science and religion. After receiving my doctorate at Catholic University I began teaching immediately in the Department of Theology at Georgetown University across town in Washington, DC. In 1970 I started offering a course on science and religion and taught it almost every year until I retired several years ago.

I have always loved science, although as an undergraduate I majored in philosophy and did my graduate work in philosophical theology. While I was teaching science and religion at Georgetown and writing books on the topic I had to do a lot of reading, especially in the areas of cosmology and biology. Although I had cut my undergraduate philosophical teeth on Thomistic metaphysics, I came to the conclusion that in spite of its brilliance Thomism could not adequately contextualize the discoveries of evolutionary biology and Big Bang physics. As the intellectual backbone of my course, therefore, I relied heavily upon science-friendly 20th-century philosophers such as Alfred North Whitehead, Michael Polanyi, Bernard Lonergan, and Hans Jonas.

My books *Science and Religion: From Conflict to Conversation* (1995) and more recently *Science and Faith: A New Introduction* (2012) reflect an approach developed over many years of teaching excellent students at Georgetown. During the 1990s I became increasingly involved in issues relating to evolution, not only because of their increasing importance in the intellectual world but also because of my frustration at the claims by creationists and prominent evolutionists alike that Darwinian science and belief in God are irreconcilable. In great measure it has been the American cultural warfare over the teaching of Intelligent Design that has led me to write such books as *God After Darwin, Deeper than Darwin*, and *Making Sense of Evolution*. These and other works have led to my lecturing often on theology and evolution both nationally and internationally. I also gladly agreed to testify on behalf of the plaintiffs at the famous Harrisburg, PA trial in 2005 against the teaching of Intelligent Design in public schools, and one of my proudest academic moments is to have received a "Friend of Darwin" award from the National Center for Science Education.

However, the main reason for my becoming interested in questions about science and religion has been my own personal search for meaning and truth. This concern has led me to embrace a vision of reality that provides ample room for both scientific inquiry and my Christian understanding of God.

2. Do you think science and religion are compatible when it comes to understanding cosmology (the origin of the universe), biology (the origin of life and of the human species), ethics, and/or the human mind (minds, brains, souls, and free will)?

Instead of talking about religion and science, it is more meaningful for me to focus here on the relationship of science to my own academic discipline, theology. I have no difficulty reconciling my Christian theological vision with scientific accounts of the universe, life, mind, and morality. I believe that everything should be open to scientific study, including human intelligence, ethical aspiration, and religion. These are, after all, natural phenomena, and everything in nature is grist for the scientific mill.

Encouraging scientists to examine everything natural does not mean, however, that science is the only fruitful way to understand nature, for there can be distinct and noncompeting levels of explanation for everything in our experience (see #3 below). For example, reading the universe as the gift of an infinitely resourceful creator does not conflict in any way with reading it simultaneously through the eyes and mind of an astrophysicist. In fact, by allowing for different reading levels we can avoid turning physics into metaphysics, or evolutionary biology into a whole worldview. What I want to avoid—as much to safeguard the integrity of science as that of theology—is the conflation of reading levels.

Such a mix-up occurs whenever a scientist or person of faith projects his or her ideological interests or worldview into what should be a purely scientific reading of nature. In this respect the marriage of evolutionary biology to materialist metaphysics is no less objectionable than the biblical literalist's reading the Bible as a source of scientific truth. A careless compounding of what should be distinct reading levels occurs especially when one reads ancient religious literature with a curiosity appropriate only to modern science. Several centuries ago Galileo rightly pointed out that theological claims will appear to conflict with science only if one first adopts the erroneous assumption that the Bible is somehow a source of scientific information. To read the Bible in this way, he observed, is to trivialize it by having it function as a mundane source of knowledge that we can acquire simply by the use of our natural faculties of observation and reason. His reflections, ignored by important church officials of his day, are similar to those expressed much earlier by St. Augustine of Hippo (354-430) and later embraced in 1893 by Pope Leo XIII who explicitly exhorted Catholics not to look for scientific information in the scriptures.

A major obstacle to adopting a plurality of reading levels, however, is the persistence of biblical literalism. As far as reading the Bible is concerned, literalism can mean different things in different periods of history. Nowadays it usually takes the form of assuming that the sacred books, if they are truly "inspired," should deliver a quality grade of *scientific* truth. This expectation ironically links some scientific materialists closely to their creationist opponents. The atheist Sam Harris, for example, insists that "the same evidentiary demands" must be used to measure the truth-status of religious propositions as those of science. So he wonders why all Christians are not as puzzled as he that the Bible, "a book written by an omniscient being" would fail to be "the richest source of mathematical insight humanity has ever known," or why the Bible has nothing to say "about electricity, or about DNA, or about the actual age and size of the universe" (*Letter to a Christian Nation*, 60-61).

In *God and the New Atheism* I give examples of how Daniel Dennett, Christopher Hitchens, and Richard Dawkins also endorse essentially the same literalism. Along with creationists the so-called New Atheists approach the Bible—since it is supposed to be "inspired"—as though it should be *scientifically* accurate. The New Atheists conclude after reading it that the Bible is not scientifically reliable, and so they can dismiss it as fiction. Meanwhile "scientific creationists" continue to interpret the biblical stories in Genesis as though these writings are scientifically superior to Darwinian biology and contemporary cosmology. My point, though, is that both creationists and New Atheists wrongly *expect* the Bible (especially Genesis) to be scientifically correct even if one side concludes that biblical "science" is sound and the other does not.

Approaching ancient texts with modern scientific expectations is the source of an unnecessary, anachronistic confusion that ends up making the Bible, and indeed the biblically based faith traditions, seem incompatible with modern science. Theology, as I understand it, however, reads the scriptures in search of levels of meaning and truth that scientific method is not wired to receive. Literalism is a way of avoiding such an encounter. It protects the fundamentalist from a serious reading of the texts while it gives the New Atheist a pretext for mocking ancient religious literature for its scientific ignorance. For example, Daniel Dennett contrasts the sorry state of scientific information in the Bible with that of present-day biology, remarking that "science has won and religion has lost," and that "Darwin's idea has banished the Book of Genesis to the limbo of quaint mythology." It seems to me that it is only because he embraces essentially the same hermeneutical expectations as the scientific creationist that Dennett can claim triumphantly that evolutionary biology has now exposed Genesis as mere illusion (interview in John Brockman, *The Third Culture*. New York: Touchstone Books, 1996, 187).

I have observed over many years that many other scientifically trained atheistic opponents of theology privilege literalist interpretations of scripture as theologically normative. They even assume it is their job to decide what should or should not pass muster as acceptable theology, complaining that theologians like me are "accommodationists" who don't count since we are not as literal minded as they expect true believers to be. Such skeptics try to convince their audiences that ancient religious texts need to be placed into a competitive relationship with contemporary science, since in that contest modern science will surely emerge victorious on the basis of "evidence." In carrying out this program they sometimes assume that all previous ages of religious and theological understanding consist mainly of awkward and now-obsolete attempts to do what science can now do better (see below). In that case, of course, it is not science but instead an anachronistic reading of ancient religious literature that makes faith and theology seem irreconcilable with science.

3. Some theorists maintain that science and religion occupy non-overlapping magisteria—i.e., that science and religion each have a legitimate magisterium, or domain of teaching authority, and these two domains do not overlap. Do you agree?

Theology and science are distinct but related ways of experiencing, understanding, and knowing. They are both rooted in the human desire to know. They both seek intelligibility and truth. They meaningfully overlap, therefore, insofar as they share a common origin in the human drive to understand, and in their both having the ultimate objective of arriving at what is intelligible and true.

However, science and theology pursue their shared goal of understanding and truth from within formally distinct *horizons* of inquiry. These horizons do not overlap, compete or conflict with each other, and what constitutes data and evidence in one is not the same as in the other. Here I am using the idea of horizon metaphorically. Visually, a horizon is the field of all things that can be seen from a definite point of view. Following Bernard Lonergan, what I mean by a "horizon of inquiry" is the whole field of things that can be experienced, understood and known from a determinate epistemic vantage point. The horizon of science consists of all that can be known from the point of view of scientific method. Obviously good scientists don't pretend to have mastered, nor do they expect ever to master, the whole field opened up by scientific inquiry. Yet they have a good sense of the kinds of questions they may or may not ask from within that horizon.

Each scientific specialty, it goes without saying, also has its own subsidiary horizon defined by the kinds of questions it includes and excludes. Nevertheless, we may speak here of scientific method in a general way as a kind of inquiry into natural causes that scrupulously avoids all considerations of purpose (final causation), value, subjectivity, importance, miracles or God.

It follows then that God—if God exists—does not show up within the horizon of scientific inquiry. One may suspect, of course, that a particular set of scientific *discoveries* (such as evolutionary biology) renders dubious the idea of a designing deity. But from within the horizon of scientific inquiry all that can be affirmed or denied are hypotheses or theories related to physical explanation. The idea of God should not even be mentioned, let alone affirmed or denied by science itself.

Why then do some people insist that science and theology are incompatible? In keeping with the metaphor introduced above, I would answer that it is because of a "horizon-fixation" originating in the scientifically unverifiable belief that the field of scientific inquiry is wide enough to embrace the totality of being. It is this horizon-fixation, and not science or theology itself, that underlies many contemporary claims, by scientific skeptics and religious believers alike, that the two are irreconcilable.

From *within* a scientific horizon of inquiry, of course, one hypothesis may compete or conflict with another. When this occurs a good rule of thumb is to choose the explanation that makes the fewest assumptions (following Ockham's Razor). On the basis of such economy and simplicity the Darwinian theory of evolution, for example, has triumphed over the Lamarckian. So meaningful conflicts often do arise within the general horizon of science, and the resolution of these conflicts often leads to impressive advances in scientific understanding.

However, there are other horizons of inquiry in addition to that of science, among which there is no meaningful competition or conflict. I believe that science, ethics, and theology are three such noncompeting horizons. They give the appearance of competing with one another only if one horizon of inquiry tries to devour the others. This can happen, for example, when a devotee of science claims that ethical values can be grounded in empirically verifiable scientific "facts," as claimed by Sam Harris in *The Moral Landscape*. Or it can happen when Richard Dawkins refers to the "God Hypothesis" in *The God Delusion*. There he tries to get his readers to agree with him that belief in God falls within the same horizon of inquiry as scientific hypotheses and that it has now been rendered obsolete by the superior—because more economical—Darwinian brand of scientific explanation.

I would agree that *within* the general horizon of scientific inquiry important conflicts may occur. But there is no good scientific reason to steal the idea of God from the field of theological inquiry and transplant it into that of scientific inquiry where it is forced to compete with other "hypotheses." It is not a redeemable part of science to hypothesize that there can be only one legitimate horizon of inquiry. Since scientific method is, in Dawkins's belief system, the only reliable way to arrive at true understanding of anything real, no other horizon is available to him within which to locate human thoughts about ultimate reality. So this is where discourse about "God" has to present and prove itself.

One can find support for this strategy in the religious world among those who follow the literalist perspective I discussed in question #2 above. My point here, however, is that Dawkins's defense of atheism is not the outcome of any rigorous application of scientific method but of an untestable belief that the modern scientific horizon of inquiry leaves room for no others. Dawkins is by no means alone. The physicist Sean Carroll seems to agree, for example, when he writes: "If and when cosmologists develop a successful scientific understanding of the origin of the universe, we will be left with a picture in which there is no place for God to act ..." (http://preposterousuniverse.com/writings/dtung/).

Theology, meanwhile, has its own horizon of inquiry. It is grounded in a qualitatively distinct set of questions from those asked by scientists and ethicists. The data that give rise to distinctively theological questions include an easily recognizable set of beliefs and ethical commitments that do not show up within the horizon of scientific inquiry, but which every scientist must embrace in order to do science at all. Here is a brief list:

1. Belief (faith or trust) that the world, including the horizon of scientific inquiry, is intelligible.

2. Belief that truth is worth seeking.

3. Belief that honesty, humility, generosity, and openness in sharing one's ideas and discoveries are unconditionally *right* (and hence that the pursuit of virtue is not irrelevant to successful scientific work).

4. Belief that one's own mind has the capacity to grasp intelligibility and to distinguish what is true from what is false.

Aware of these four grounding assumptions, theology asks questions about them, and these questions generate a horizon of inquiry logically distinct from that of science or ethics:

1. Why is the universe intelligible at all?

2. Why is truth worth seeking?

3. What makes my commitment to honesty (and other virtues essential to being a good scientist) unconditionally right?

4. Why should I trust that my mind, in spite of all its limitations, can still grasp intelligibility and make true judgments?

These "limit-questions," as theologian David Tracy calls them, do not arise as the result of the accumulation of information within the heuristic territory explored by science, nor are they merely the product of growth in ethical sensitivity. Rather they are the result of a *horizon-shift* that can take place when a reflective knower realizes that there are kinds of questions that cannot be meaningfully dealt with from within the limits of scientific and ethical inquiry.

Although I don't have the space to expand on it here, to me the most satisfying answer to these limit-questions is a theological one, namely, that the universe—along with the minds and consciences that have emerged from it—is ultimately enveloped and sustained by an ultimate horizon of infinite Wisdom, Truth and Goodness, names that Christian theology has used since antiquity when referring to God as the ground, source and destiny of all finite being. This horizon grasps us more than we grasp it.

No doubt, some opponents of what I've just said will insist that my limit-questions are not meaningful questions at all. My response is that I fully agree that from within the horizon of scientific inquiry they are not meaningful—for reasons I have already given. But perhaps my critics will notice that in their attempts to show that I must be wrong they too have slipped beyond the limits of any purely scientific field of exploration and are being guided perforce by the very same four-fold set of commitments they are now seeking to debunk. These commitments in turn raise questions that open up a horizon of inquiry qualitatively distinct from that of scientific method. Theology, of course, is not the only occupant of this distinct horizon, since any number of alternative worldviews or metaphysical experiments, including a pure naturalism or materialism, may be tried out there. All I am saying is that while theology may conflict meaningfully with these other worldviews, this clash of theology with other worldviews in no way entails a conflict of theology with science.

4. What do you consider to be your own most important contribution(s) to theorizing about science and religion?

My contributions lie mainly in the area of the theology of nature and especially the theology of evolution (*God After Darwin* and *Christianity and Science*). Theology is a discipline that reflects systematically on the meaning of faith from within the limited perspective of a religious tradition, in my case Catholic Christianity. Theology does not look at the world from some imagined universal perspective that ignores differences in the world of religions. As a theologian interested in science and its relationship to faith, therefore, my first task as I approach such topics as the meaning of evolution must be to bring to the fore what the word "God" means in my tradition. For Christians what "God" means is inseparable from the picture of Jesus presented in the Gospels and interpreted by other biblical and classic Christian texts and traditions throughout the centuries. The constant teaching of Catholic tradition, ratified by numerous councils and creeds, is that Jesus is the normative manifestation in human history of the eternal mystery of God. Catholic tradition instructs us never to take our eyes and minds off of the image of Jesus when we think and talk about God or when we worship along with others.

One of the themes suggested by theological reflection on the picture of Jesus is that of God's humility or "descent," that is, God's decision to live like a creature, to participate fully with humans and other living beings in their struggles, suffering and death, and thus to open creation to a new and redemptive future. The image of a God whose very nature is self-giving love supplants all ideas of a divine potentate or engineer that people have projected onto "God." When I think about the issue of evolution and theology, therefore, I do not try to reconcile Darwinian science with the "hypothesis" of divine design. I do not ask whether design points to deity, but whether the long story of life and the universe carries a meaning. As Pope John Paul II testified in his encyclical *Fides et Ratio* (1998), the God of Catholic faith and theology is one who undergoes a *kenosis* (that is, a pouring out) of the divine substance in self-giving love to the creation. This understanding of God, articulated especially well by the 20th Century Catholic theologian Karl Rahner, comes from centuries of prayerful meditation on the self-sacrificial obedience of Jesus who is called the Christ. Hence it is the chief purpose of Catholic systematic theology, as John Paul II insisted, to highlight the divine *kenosis*. The kenotic descent of God, he wrote, is a "grand and mysterious truth for the human mind, which finds it inconceivable that suffering and death can express a love which gives itself and seeks nothing in return" (# 93).

The point I am trying to make here, then, is simply that I am not obliged to locate my thoughts about God within a purely scientific horizon of inquiry. The idea of a self-emptying God fits no conventional conceptual scheme devised by human speculation, least of all one in which the idea of God is a quasi-scientific hypothesis that has now been rendered obsolete by the arrival of modern science. The image of God that informs Christian theology opens up a whole new horizon for looking at the universe and the life-process. Acknowledging the distinct horizons of science and theology allows me to embrace the enigma of evolution without requiring that I ignore, modify or slant in the slightest way the data gathered by such fields as geology, biology, paleontology, anthropology, biogeography, comparative anatomy, and genetics. Finally, after consenting to the plurality of distinct horizons, it is the mission of a theology of nature to integrate them into a synthetic vision wherein differences do not dissolve but instead contribute in distinct ways to the larger and longer human quest for meaning and truth. This synthesis is what I have attempted in the books mentioned above.

5. What are the most important open questions, problems, or challenges confronting the relationship between science and religion, and what are the prospects for progress?

The fundamental challenge is to have the various sides in the discussion come to an agreement about the differences between scientific method on the one hand and the various ideological interests and worldviews that have been unfortunately alloyed with science on the other. The conflation of science with materialist philosophy, for example, is no less corruptive of intellectual integrity than is the reading of biblical creation stories as though they are trafficking in scientific truth. Such contamination is an obstacle both to the prospering of science and to a fair and honest interpretation of religious literature.

Instances of this intellectual impurity show up these days not only in the popular writings and lectures of some well known evolutionists but also in those of contemporary scientists and philosophers who fuse the horizon of physics with that of metaphysics, and of cosmology with ontology. This (con)fusion occurs glaringly in claims that physics may now replace theology in explaining how the universe came into existence out of "nothing." Once again, however, such a proposal could be taken seriously only if there were good reasons to assume that the metaphysical territory formerly occupied by theology can now be meaningfully merged with the province of empirical science.

No assumption, however, would be more corrosive of the integrity of science or more misinformed about the nature of theological inquiry. Here I must limit myself simply to identifying this confusion as a great obstacle to true progress in conversations in science and religion. I am thinking, for example, of how the words "nothing" and "nothingness" get tossed around by some physicists today when they can legitimately be referring only to subtle transformations that occur *within* the natural world. Science by definition can have no other horizon than the natural world—which after all is something and not "nothing." Empirical science in fact can make contact neither with the supernatural nor with an abyss of ontological nothingness. Referring to a universe-generating quantum vacuum as "nothing," for example, has absolutely nothing to do with "nothing" as the term is used in the theological doctrine of divine creation of the world *ex nihilo*. Such confusion can be avoided only by tolerating a plurality of explanatory levels and distinct horizons of inquiry.

12

Muzaffar Iqbal

Muzaffar Iqbal is the founder-president of the Center for Islamic Sciences, Canada; editor of *Islamic Sciences*, a semi-annual journal of Islamic perspectives on science and civilization; and General Editor of the seven-volume *Integrated Encyclopedia of the Qur'an*. Dr. Iqbal received his Ph.D. in chemistry from the University of Saskatchewan, and then left the field of experimental science to fully devote himself to study Islam, its spiritual, intellectual, and scientific traditions. Born in Lohore, Pakistan, he has lived in Canada since 1979. He has held academic and research positions at the University of Saskatchewan, University of Wisconsin-Madison, and McGill University. During 1990-1999, he lived and worked in Pakistan, first at the Organization of Islamic Conference Standing Committee on Science and Technology (COMSTECH) and then at the Pakistan Academy of Sciences. Dr. Iqbal has written, translated, and/or edited twenty-one books and published nearly one hundred papers on various aspects of Islam, its spiritual and intellectual traditions, and on the relationship between Islam and science, Islam and the West, the contemporary situation of Muslims, and the history of Islamic science. His books include *Islam and Science* (2002), *Science and Islam* (2007), *Islam, Science, Muslims, and Technology: Seyyed Hossein Nasr in Conversation with Muzaffar Iqbal* (2007), *Dew on Sunburnt Roses and other Quantum Notes* (2007), *Definitive Encounters: Islam, Muslims, and the West* (2008), and *The Making of Islamic Science* (2009).

---

1. What initially drew you to theorizing about science and religion?

The dawn of the fifteenth Islamic century (November 1979) brought an enormous amount of spiritual and intellectual energy to the Muslim world. The following decade produced world-changing events (the Islamic Revolution in Iran, the Soviet occupation of Afghanistan and the Iran-Iraq War, civil wars in Lebanon and Sudan, the First Intifada, OPEC control of rising oil prices) and also brought a chilling fact to the attention of Muslim intellectuals: during the preceding three centuries, all branches of knowledge had undergone a fundamental shift in their foundational structure and the resultant epistemological framework had all but eliminated the Divine from the equation. While this secularization of knowledge is a matter of great concern for all religious traditions, it had enormous consequences for Islam, which had long fostered scholastic and scientific inquiry rooted in religious sources. Alarmed by this state of affairs, Muslim intellectuals called for the "Islamization of knowledge," a project that would re-examine the modern humanities and social sciences in the light of Islamic principles.

In the social and political context of the 1980s, science flourished in not a single Muslim country, despite their enormous resources and new oil wealth. This science "deficiency" led to an insatiable hunger, yielding unrestrained calls by multifarious leaders for the acquisition of modern science—a call that was often clothed in religious dress, perhaps to gain legitimacy and to inspire a younger generation. At that time, there was little realization by these leaders that science had become the final arbiter of truth-claims. This was, indeed, an unprecedented development: at no time in history had a single branch of knowledge attained such uncontestable authority. Both in public perception and in the academic world, higher-order questions (the origin of the cosmos and its ultimate fate, life on earth) were deferred to the hegemonic power of science to unveil the true nature of things.

Distinct schools emerged from the confluence of these two factors (the Muslim recognition of the secularization of modern knowledge and the dearth of Muslim scientific inquiry). One school held that Islam itself is altogether antithetical to scientific inquiry (and hence implicitly urged further secularization in order to secure the conditions for science to flourish). When faced with the historical scientific legacy of Islamic civilization (which, after all, gave the world the scientific tradition with the greatest longevity to date, lasting from the 8th to the 16th century), this school claimed that the Muslim world was merely a kind of depot for Greek science, which it received through the Translation Movement (9th and 10th centuries) sponsored through the patronage of a few exceptionally enlightened Caliphs and their courtiers. The scientific knowledge thus acquired was then passed on to Europe, its rightful heir—while Islam itself was still held to vigorously oppose such inquiry.

Another school took the opposite stance, arguing that Islam and Islamic civilization prefigured and anteceded every modern scientific effort. This approach also generated an offshoot: an enormously popular trend that links all manner of scientific discoveries to the Qur'an: even Darwin has been found to have Muslim precedents! This approach (not even a school proper—an absolutist hermeneutic) superimposed Qur'anic verses over modern scientific data and claimed religious sanctity for all kinds of modern scientific theories.

Meanwhile, having completed my PhD in chemistry from the University of Saskatchewan, I had personally been dismayed by the failure of science to provide answers to my "ultimate questions." I understood the structure of the atoms that constitute our cosmos, but the cosmos itself remained a mystery. I understood the biological systems that govern life, but life itself eluded these systems. In the summer of 1986, I was in the sub-sub-basement of the Neurological Institute of McGill University smashing fluorine atoms in a cyclotron to extract concentrated quantities of their radio-active isotope, which was in turn used to develop new drugs to trace brain tumors—but my own spiritual quest remained unrequited and the inner fissure became increasingly unbearable. This was not merely intellectual curiosity; it was a deep, existential crisis, requiring immediate and real answers.

My search had already drawn me deep into literature (by then, I had plumbed both my own literary tradition as well as Western traditions, including the Latin American and Russian literary giants; I had written a number of short stories and poems and my first novel was published in 1987), history, philosophy, and mysticism. Yet I was conscious of the deep chasm produced by these various opposing currents.

When some resolution did emerge for me in the winter of 1988 in Fort Chipewyan—a remote and geographically isolated town in northern Alberta—it was sheer Divine mercy. That spiritual change produced an intellectual clarity and discursive thought started to put things in communicable and coherent arguments. Shortly thereafter, I moved to Pakistan, where I lived and worked in Islamabad (1990-1999). Among other things, I was there the editor of *Islamic Thought and Scientific Creativity*, a quarterly journal devoted to the exploration of science in the Islamic polity. It was during this period that I started to theorize religion and science in a more formal way.

In the mid-1990s, when science and religion dialogue surged in America—and to some extent in Europe—I was naturally drawn to it, only to discover that the entire discourse, its key questions and terms, had been framed around *one particular* religion (Christianity) and one particular scientific tradition (modern science with its particular European history). This framework adopted by a much broader, international, multireligious discourse (said to address science and religion writ large) led me to re-examine its fundamental parameters. Religion is a vacuous term, until we fill it with content from one particular religious tradition; likewise, science cannot be reduced to a singular mode of enquiry. True, Chinese, Greek, Roman, and Islamic scientific traditions today flourish nowhere, but this does not mean that "science" in the broader academic field of science and religion should be understood merely as modern science. That already accepts certain definitions, contexts, narratives... In the long history of science, modern science is the latest entrant; there is no guarantee that this particular epistemology will continue to prevail. The field ought not to limit itself by presupposing in its analytic apparatus a particular, modern rendering of these concepts. But of course, the conditions that provoked the rise of science-and-religion discourse in America have something to do with this.

2. Do you think science and religion are compatible when it comes to understanding cosmology (the origin of the universe), biology (the origin of life and of the human species), ethics, and/or the human mind (minds, brains, souls, and free will)?

In the context of what I have just said, it is obvious that the answer to this question depends on *which religion* and *which cosmology* (or biology, ethics, human mind) we are talking about. Also, the answer would depend on what is meant by "compatibility." Let me briefly explain. Science, taken as accumulated and systematic human knowledge about the physical world, is the universal heritage of all humanity, but each civilization has historically produced its own peculiar epistemology of this inquiry and every religious tradition understands the physical world in its peculiar way, even though there might be certain common domains. Thus, since there are distinctive ways in which science, as a civilizational activity, is understood in different traditions, we need to explore the question of compatibility within each religious and scientific tradition. In other words, while raw data remains the same whether one conducts an experiment in Makkah or Washington, DC, there are different interpretive tools which infer different results from the same data. Modern science has enabled us to accumulate a tremendous amount of new data about great cosmic regularities (the precise orbits of planets, constant parameters of light and other forms of energy, etc.) and neurological patterns and all manner of other things. Yet this scientific data does not speak for itself. It needs interpretation, whether as divinely ordained or self-sustaining or otherwise. The data itself is not caught between science and religion; it admits the presuppositions of a host of theoretical frameworks.

Likewise, while there is a certain common domain held by religious traditions, one cannot really speak of "religion" as a generic term, without specificity: what is understood by the term "religion" by a Muslim is vastly different from what it means for a Hindu or a Buddhist, or even for a Christian or a Jew. Views on divine agency, human volition, and the human mind and body differ widely within and across these traditions—all of these form the science and religion nexus. Thus, given the enormous variety of religions of humanity as well as the large number of scientific and supra-scientific theories about the origin of the cosmos, the nature of life, and human volition, one cannot say anything meaningful without being specific. If we understand contemporary scientific theories to be monolithic and likewise circumscribe religion as one set of beliefs or another, then we can continue to churn out "yes" and "no" answers. If, on the other hand, we understand scientific truth claims about these ultimate questions to be a work in progress and delve deeper into the religious teachings about origins in each of our religions—only then do we enter into a meaningful discourse on these important questions.

I think there are numerous "science and religion discourses" at this point of time; every religious tradition is engaged with a discourse on the nature of cosmic and human realities and mysteries as new scientific data brings new facets to light. Although now we have only one "science" left, one cannot allow this monolithic understanding to stand in absolute authority. It is possible that five hundred years from now, a new science would emerge, that is, a new way of studying the physical world, with a distinctive epistemology, operative methodology, a new sociology, and a new economics of science, perhaps not wedded to technology and the market economy as the present science has become.

Speaking from my own religious perspective, I cannot but marvel at the enormously rich religious discourse on cosmology, origin of life, ethics, and the extent of human free will that for centuries inspired and informed the Islamic scientific tradition. Likewise, many religious scholars produced works in various branches of science, such as astronomy, medicine, mathematics, and physics. Ultimate truth claims about cosmology, biology, ethics, and human mind remain—as they are likely to—in the realm of modern scientific uncertainty. Meanwhile, religious traditions approach these questions with an alternate set of coordinates, by which there is often an authoritative arbiter (a revealed book, a divinely-appointed human, etc.). From the perspective of religious quest, it is the human understanding of these answers that produces discursive thought, explicating how these truth claims are understood.

In the final analysis, religious cosmology rests on a Divine existentiating command: creation ex nihilo, the Qur'anic *kun* ("be!"). While modern science theorizes about these ultimate questions, it can only offer data and theories of varying degrees of reliability after the initial Bang (big or small), after a sperm and ova have been infused to produce life—but the mysteries themselves of cosmos and of life remain beyond the ken of science.

In a certain sense, one can only talk of compatibility or its lack in our religious and scientific understandings of the cosmos and the human condition *after* the initial conditions are secured (that is, either by suspending one's inquiry into initial conditions, or settling for one of the several available hypotheses).

In addition, there are specific questions within each religion and science discourse. For instance, in the Islamic tradition, the question of compatibility of Islam and science is not today the primary question; rather, it is the matter of the total divorce between the religious beliefs of a Muslim scientist and the science he or she is practicing that is of fundamental importance. This radical separation has occurred through a historical process which has yet to be fully understood; nevertheless, for Muslim scientists, the vast dissonance between their religious worldview and the contemporary scientific worldview (which they themselves have had little part in constructing) remains a seemingly unbridgeable chasm. I would add that, at least in certain domains of inquiry, it would be the same for a Christian and a Jewish scientist: no religious tradition can ignore the deadly consequences of the Faustian pact between the contemporary science and technology on the one hand and technology and the market economy on the other, consequences evident in the destruction of countless species, desecrated forests, emergence of weapons of mass destruction, pollution of the food chain, and the environmental crisis which now looms so large that no one can ignore it.

3. Some theorists maintain that science and religion occupy non-overlapping magisteria—i.e., that science and religion each have a legitimate magisterium, or domain of teaching authority, and these two domains do not overlap. Do you agree?

Not completely. All religious traditions do interact with scientific data through various possible logical relations, including both marginally overlapping as well as non-overlapping zones between them. Seen from an Islamic perspective, religious beliefs are ultimately and deeply rooted in an existing reality, only some parts of which are accessible to humans. In other words, the domains of existence revealed by revelation exceed those to which human beings have access. Yet a large part of the Qur'an deals with those realms of the existence to which humans do have access: in the typology of the classical exegete Fakhr al-Din al-Razi, these comprise the physical world, human history, and marvelous events within the human soul. These three domains, which are to a large extent within human reach, are explored through natural and human sciences; and they are also the subject matter of religious discourse. Although the hard data yielded by these sciences and disciplines is independent of religious beliefs, interpretation of that data is necessarily subjected to and framed by those beliefs. Water freezes at zero degrees and boils at one hundred (under a certain atmospheric pressure, of course). But why does it do so? As soon as we ask the "why" question, we enter into overlapping magisteria: science attempts to explain this on the basis of weak hydrogen bonding between oxygen and hydrogen, the two atoms that constitute water; religion provides a teleological explanation. Are they compatible? Indeed, they are, but only through a particular epistemological framework that integrates various explanations at different orders. This framework is what today demands explicit attention and articulation, being polemically contested by both scientific and religious factions.

In addition, we must add that contemporary religious discourse in Islam is greatly truncated due to the historical circumstances which have given birth to the rise and expansion of modern science and technology—a process in which Muslims have had little role, which again has to do with the global market economy and the legacy of colonial centres. In a certain way, many religious scholars today are working under a deep inferiority complex with regard to science while others defensively react against it. Thus one sees either wholesale condemnation of science or blind submission. These two extremes then inform the so-called "Islam and science" discourse in various ways and lead to erroneous conclusions about their mutual compatibility or its lack. Other religious traditions would have their own internal mechanisms and specific conditions which would have a decisive impact on the answer to this question.

4. What do you consider to be your own most important contribution(s) to theorizing about science and religion?

If I can claim anything in this regard, it is some sustained work on two fronts: critical input into certain fundamental aspects of the religion and science dialogue, insisting that one cannot talk of these issues in a monolithic way but rather needs to examine each religious tradition in its own specificity (for which see my *Science and Islam* [Greenwood 2007], reprinted with afterword as *The Making of Islamic Science* [Islamic Book Trust 2009]; also my *Islam, Science, Muslims, and Technology: Seyyed Hossein Nasr in Conversation with Muzaffar Iqbal* [al-Qalam Publishing and Islamic Book Trust, 2007; Iranian ed. 2008, Pakistani ed. 2009]) and my modest contribution to Islam and science discourse within the context of history of science (for which see my *Islam and Science* [Ashgate 2002, Pakistani ed. 2004]; my co-edited volume *God, Life and the Cosmos: Christian and Muslim Perspectives* [Ashgate 2002, Indonesian ed. 2006]; various issues of the semi-annual journal I edit, *Journal of Islam & Science* [2003–present]); and the work of our Center (Center for Islamic Sciences) over the last fifteen years.

Most recently, I edited a four-volume (2000-page) anthology that seeks to present an integrated view of Islam and science discourse from various perspectives, bringing together works from recent studies in history of science, philosophy of science, exegetical studies, and other texts based on Islamic view of the cosmos. Intended as a comprehensive reference resource and titled *Islam and Science: Historic and Contemporary Perspectives* (Ashgate 2012), the anthology brings together the most important and influential articles dealing with various aspects of the relationship between Islam and science, covering a wide spectrum of subjects and topics (from Islamic perspectives on cosmology and biological evolution to the relationship between the Qur'an and science).

After tracing certain historical currents that shape encounters between modern scientific and Islamic religious traditions, I began to notice patterns in the ways these encounters are framed, provoked, addressed, and (sometimes) resolved. These patterns often have less to do with the substantive issue at stake (whether the teaching of evolution or bioethical questions related to organ donation or end-of-life care) than with contemporary politics and conflicting interpretive principles. My research over the past decade has thus shifted toward hermeneutical questions in Islamic scientific tradition (science as organized knowledge), from the modes of scientific enquiry to their paradigmatic case, the Muslim traditions of Qur'an interpretation.

5. What are the most important open questions, problems, or challenges confronting the relationship between science and religion, and what are the prospects for progress?

The open questions include those regarding consciousness and human perception of reality, both from scientific and religious perspectives; possibilities for science to more productively engage with meta-scientific questions, namely those regarding the origin and ultimate end of the cosmos; and questions related to mysteries of life and miracles (miracles defined as exceptional, supra-rational events). One cannot but remain hopeful that ultimately there will be some kind of realization that science is but one tool that humanity has to understand these questions and that this tool cannot be used in isolation of other means of perception and understanding of reality.

In addition to these general (and big) questions, and as noted above, each "religion and science" discourse faces questions unique to the specific traditions engaged. For the Islamic tradition, one of the most important open questions is how Muslims can remain faithful to their fundamental beliefs about God, life, and the cosmos as historically imparted and debated while participating in a science that operates on the fundamental assumption that the object of its study (the physical world) has no Divine sanctity, that it is just matter, which can be manipulated for perceived benefits to humanity. The physical world, as understood from a religious perspective, is a living entity, cognizant of its Creator, responsive to His commands, made subservient to humans yet *with certain conditions*, the greatest of which is to be its custodians (*khulafa*), that is, to utilize its resources with a deep sense of thankfulness and with utmost responsibility.

In addition, there are numerous further questions arising from the intimate relationship between modern science and modern technology. The latter has not only allowed humans to go to distant planets, it is also prompting redefinitions of fundamental religious precepts involved in anthropology and psychology. For instance, in the Islamic tradition, the human person is approached through a set of concepts (*insan, bashar, nafs, ruh, khalq, sura...*) each related to an aspect of the relation with the Divine, whether the creation of the first human in the image of God (with a range of ethical and mystical interpretations of what this means) to the ennobling of humanity through the Divine impartation of innate knowledge, God teaching the human the names of all things. Furthermore, as various technologies reshape daily lives by changing to processes related to the procurement of food, modes of transport and communication, and the like, Muslims face the mighty task of understanding their religious practices in the context of this technologically-construed world, leading to the violent rupture of centuries-old cultural practices and norms.

13

Lawrence Krauss

Lawrence Krauss is an internationally known theoretical physicist and cosmologist. His work focuses on the interface between elementary particle physics and cosmology—which includes research on the early universe, the nature of dark matter, general relativity, and neutrino astrophysics. He is currently Foundation Professor of the School of Earth and Space Exploration and Inaugural Director of the Origins Project at Arizona State University. He is the author of over 300 scientific publications as well as numerous popular articles on physics and astronomy. He is also the author of several popular books including *The Fifth Essence: The Search for Dark Matter in the Universe* (1989), *The Physics of Star Trek* (1995), *Beyond Star Trek* (1997), *Atom: An Odyssey from the Big Bang to Life on Earth...and Beyond* (2001), and *A Universe from Nothing: Why There is Something Rather than Nothing* (2012). Professor Krauss is the recipient of numerous awards for his research and writing, and is the only physicist to have received awards from all three major U.S. physics societies—the American Physical Society, the American Association of Physics Teachers, and the American Institute of Physics. In 2012 he was awarded the National Science Board's Public Service Medal for his contributions to public education in science and engineering in the U.S.

---

1. What initially drew you to theorizing about science and religion?[1]

I don't know whether theorizing is the right word, but what drew me to begin to think about the issue was when I lived in Ohio. There was an effort to add creationism to the science biology course in addition to evolution. What surprised me at the time was that the biology community wasn't as outspoken as I would've liked. For better or worse, I had a reasonably high public profile, and I felt since this was a question of biology, this was a question of science. I wanted to defend science. So, I began to speak out and be active in terms of countering these people who somehow thought creationism was science and were trying to distort and dilute the teaching of science in schools—which I think is a travesty and a crime. That got me started on this. Of course, once one gets started, people begin to come to you more and more, especially if you have a reasonable public profile. So I got involved in it around the country.

My thoughts have evolved, because I had previously spoken about the fact that you certainly can't disprove the existence of God or the

---

[1] This interview was conducted via Skype and the transcription approved by Lawrence Krauss. Any remaining errors or mistakes are the fault of the editor.

existence of some vague purpose in the universe. But a colleague of mine pointed out to me that I was being a little disingenuous because the dogmas of the world's organized religions, based on their holy books, are in fact all inconsistent with science. To pretend they aren't is to mislead people. I guess after I was reminded of that fact, I changed my tone a little bit.

Then, the more I got involved, the more I began to see that in many cases religion was the source of an anti-science mentality that I think is extremely dangerous. I think, for me, it's not so much fighting religion as promoting the beauty and wonder of the universe as we understand it through science. That's what I want to do, but I also want to work against those things that get in the way of that.

2. Do you think science and religion are compatible when it comes to understanding cosmology (the origin of the universe), biology (the origin of life and of the human species), ethics, and/or the human mind (minds, brains, souls, and free will)?

Absolutely not. If you mean by "religion" organized religions, they are all manifestly inconsistent with what we know to be true about the universe. There's no room for compatibility between nonsense and sense as derived by empirical evidence.

Religion, when it isn't wrong—which *it is*, manifestly, through the scriptures of the different religions by disagreeing with what we know about the universe—when it isn't wrong, it simply doesn't add anything. It assumes the answers before it asks the questions. It's based on, in most cases, the writings of Iron Age peasants who didn't even know the earth orbited the sun. Why would we expect those things to add any insight in a modern world of understanding fundamentals of cosmology, biology, or the human mind which are the forefront of our modern science and which have taken hundreds of years to learn about? So, absolutely, there is zero compatibility.

Humans don't live inside whales. The Sun doesn't go around the Earth. The Sun cannot stop in the sky as a result. These silly miracles in the scriptures are obvious nonsense. But more deeply, there is no evidence of any divine creation in any of the objects we measure in the universe, and we can understand how the atoms in our bodies can arise from nothing without violating laws of physics. As for ethics and morality, much of what passes for morality in the Bible is obscene, from the rampant genocide in the Old Testament to the disgusting notion in the New Testament that making a decision to not follow Jesus will condemn you to eternal torture.

3. Some theorists maintain that science and religion occupy non-overlapping magisteria—i.e., that science and religion each have a legitimate magisterium, or domain of teaching authority, and these two domains do not overlap. Do you agree?

Absolutely not. I don't know what those non-overlapping domains are. Some people say that science is about what is and religion is about what *ought* to be, but that's ridiculous. Because you don't know what *ought* until you know what *is*, first of all. Secondly, I don't think religion has any authority in the domain of what ought.

As I have said, I can't think of more immoral, if you want to use the word—and I don't like to use the word immoral but I'll use it—writings than those in, say, the Old Testament and, to some extent, the New Testament. Religion has no authority except the authority it establishes for itself, where science uses nature to determine the answers to questions. Religion has no 'magisteria,' as far as I can see, except for closing minds and distorting reality.

4. What do you consider to be your own most important contribution(s) to theorizing about science and religion?

I don't know if I've made any important contributions. But I guess I would say my efforts, the most important ones, have been to point out that promoting nonsense is really an attack on science at a fundamental level.

You don't have to be a biologist to defend the teaching of evolution, because evolution is an idea that's been tested and is at the basis of modern biology. You don't have to be an expert in that.

I think probably my most recent book—*A Universe from Nothing: Why There is Something Rather than Nothing* (2012)—in some sense has had probably its biggest impact by pointing out we long ago got rid of God when it came to biology. We got rid of special creation. We got rid of all of that. Darwin did that. So, religions have pushed the question back, not from where did life come from, but where does the universe come from. I think by pointing out that everything we know is consistent with a universe that came from nothing without any supernatural shenanigans has had, as far as I can see, a rather big impact. Maybe that's my most important contribution.

5. What are the most important open questions, problems, or challenges confronting the relationship between science and religion, and what are the prospects for progress?

Well, I don't think there are any important questions. There is no relationship between science and religion, as far as I can see, and pretending there is just demeans science and elevates religion.

I guess if I were to look at the next areas where I think there'll be the most interesting developments that confront religious ideas, they might arise in neuroscience. When I think about the area of neuroscience we start to understand in more detail the nature of consciousness and issues like morality in a context of understanding how the brain works, that that will again force a re-definition and a new understanding of concepts which previously were assigned to some non-existent deity. So I think probably the area where there will be the most development, in terms of confrontation, will probably be in the area of neuroscience and our understanding of mind.

Lastly, I'll add that I'm heartened that as far as I can tell, in the first world, religious affiliation is monotonically decreasing. Hopefully, one day it will go to zero.

14

Colin McGinn

Colin McGinn is a world-renowned British philosopher best known for his work in the philosophy of mind. He has held teaching posts and professorships at University College London, the University of Oxford, Rutgers University, and the University of Miami. McGinn is the author of dozens of scholarly articles and over twenty books, including *The Character of Mind* (1982), *Mental Content* (1989), *The Problem of Consciousness* (1991), *Knowledge and Reality: Selected Papers* (1999), *The Mysterious Flame: Conscious Minds in a Material World* (1999), *The Making of a Philosopher: My Journey Through Twentieth-Century Philosophy* (2002), *Consciousness and Its Objects* (2004), *Mindsight: Image, Dream, Meaning* (2004), *The Meaning of Disgust* (2011), and *Basic Structures of Reality: Essays in Meta-Physics* (2011).

---

1. What initially drew you to theorizing about science and religion?

If by "science" we mean the physical sciences (physics, astronomy, chemistry), the biological sciences (genetics, zoology, ethology), and the social sciences (psychology, sociology, economics), then I have not much been drawn to think about the relationship between science, as so defined, and religion. But if we include philosophy as a science (which I believe we should), then I have considered the relationship between philosophical science and religion. Or better, I have been interested in the relationship between religion and rational thought, construed broadly. Science, in the narrow sense, constitutes only a subclass of rational thought—which includes, in addition to philosophy, logic, mathematics, literary criticism, history, linguistics, animal husbandry, and many other things. The relevant distinction in all this is between whether the discipline in question relies on rational methods of enquiry and evaluation, on the one hand, and whether it relies on faith and superstition, on the other. The real debate is between faith and reason, not religion and science. Science is just one form of reason; ethical thought (for example) is another exercise of reason. Faith is an exercise of *non*-reason. So I have been concerned with where reason leaves off and faith begins, and whether beliefs based on anything other than reasons can be acceptable. I have concluded that they cannot, despite the breadth and variety of reason. In short, the problem with religion is not that it is unscientific (though that can be a problem); the problem is that it is *irrational* (and proudly so). There may be psychological "leaps of faith," but the trouble is that they never land you with a respectable belief that others can be expected to share.

The real conflict, for me, then, is between rational philosophy and irrational religion. Most of my objections to religion are philosophical in nature, not science-based. For instance, I reject the idea that ethics can be founded on divine commands, for the reason given by Socrates long ago—namely, that moral rules are not right because God commands them but rather God commands them because they are already right. This is the famous "Euthyphro argument" and it has nothing to do with science, but rather with careful logical analysis. For another instance, I don't think the notion of a purely "spiritual" disembodied thinking being makes any metaphysical sense. Nor do I think the traditional philosophical arguments for the existence of God are logically sound. For me it is all about logic, not empirical science.

2. Do you think science and religion are compatible when it comes to understanding cosmology (the origin of the universe), biology (the origin of life and of the human species), ethics, and/or the human mind (minds, brains, souls, and free will)?

Traditional religions combine factual and prescriptive claims. The factual claims can concern matters of ascertainable empirical fact or they can be metaphysical in nature. The former are the domain of science: thus the origin of the material universe and the origin of life. Religions tend to incorporate the speculations of earlier times, often going back thousands of years, and have little or no empirical support—they are just naïve guesses as to how things came about. Modern science has refuted many of these guesses—as with big bang cosmology and Darwin's theory of evolution. In so far as a religion denies any of these scientific findings, it is incompatible with established science and must be rejected. Religious thought may contrive ways to keep God in the picture in these originative events (lurking in the background somehow) but if it actually denies the scientific accounts it is empirically false. For example, if religion denies that all animals are related—holding that each species was created separately—then we know this to be empirically false; similarly, for the origin of humans in early ape-like ancestors. Or if religion maintains that the galaxies, stars and planets, all came into existence simultaneously, not deriving from a previous gaseous state, then this too is known to be empirically false. This should not be surprising, given that most religions incorporate early speculations about matters of fact that were hard to investigate at the time. Knowledge progresses, but religions tend to cling to ancient and outmoded doctrines.

As to ethics, here we enter a different domain. The subject of normative ethics (what is right and wrong, and why) is not a scientific subject at all, in the narrow sense of "science" we are operating with. It concerns *rational thought* in the domain of value, not empirical thought in the domain of fact. Strictly speaking, science has nothing to say about ethics in this sense (though it may have something to say about how we form ethical beliefs and other attitudes). The dispute between, say, consequentialists and deontologists, say, is not going to be settled by experimental means. Religion, however, with its prescriptive component, does bear directly on ethics. The question, then, is whether its ethical prescriptions comport with the most rational ethical principles we can construct or discover; and here the potential conflict between religion and philosophical ethics arises. Without going into detail, the general point to be made is that religious ethical precepts may or may not be seen to be justified in the light of critical rational reflection—but they are *subject* to such reflection. They are not immune to criticism simply because they emanate from some religious tradition; they must face the tribunal of rationality too. They are not "protected" from rational scrutiny because they wear the badge of religion and claim divine origin; they are as open to rejection as any other ethical prescription. If, for example, they prohibit certain actions and relationships in such a way as to cause great suffering, then they may be faulted on utilitarian grounds; or if they recommend violence against non-believers. Treating animals with disregard for their rights and interests would be another example.

On the question of free will, again the question is philosophical: if a religion insists on the reality of free will, but a convincing philosophical argument defeats free will, then the only rational response is to reject free will. This is not a scientific issue but a philosophical one. Religions often adopt positions on philosophical questions, as well as on scientific ones, but they may take the wrong position (another example is the immortality of the soul). What is objectionable is accepting positions as a matter of "faith" or "divine revelation" or "scripture," instead of reasoned argument.

3. Some theorists maintain that science and religion occupy non-overlapping magisteria—i.e., that science and religion each have a legitimate magisterium, or domain of teaching authority, and these two domains do not overlap. Do you agree?

I have already answered this question, but to repeat: religions typically make empirical, philosophical, and prescriptive claims. All these claims can come into conflict with superior forms of the same types of claims—from science, from philosophy, and from a more sophisticated ethics. There is much overlap of these kinds, and therefore much error in traditional religious doctrines (though sometimes the doctrines are sensible enough, as with a fair amount of religious morality). The doctrines of religion simply reflect what people thought at earlier times, sometimes going back a long way—but science, philosophy, and ethics have made progress since ancient times. The point is that religion offers no privileged source of knowledge, in any domain; it is as vulnerable to rational scrutiny as any body of human opinion inherited from earlier generations. The problem is that religion tends to exempt itself from rational evaluation, falling back on notions of "faith" and "revelation"— but these are not genuine sources of knowledge. Only reason can make beliefs reasonable, and beliefs must be held reasonably or not at all.

A separate question is whether science and philosophy (including ethics) are "non-overlapping magisteria," and here I think the answer is affirmative. This is a big subject and I can't treat it fully here, but I believe (following a long tradition) that science and philosophy ask different kinds of questions and use different methods. In short, science is a form of empirical inquiry, while philosophy consists of conceptual analysis. Science is a posteriori and philosophy is a priori. This means that science cannot answer properly philosophical questions, which are largely definitional and logical. Science cannot tell us whether skepticism is true or whether free will is compatible with determinism or whether the mind is just the body or whether the only good is happiness—these are distinctively philosophical questions. So science does not exhaust the field of rational human inquiry, not because there are specifically religious questions it is not equipped to answer, but rather because there are specifically philosophical questions it is not equipped to answer (and some of these questions are taken up in religious doctrine). What I deny is that there are any (genuine, interesting) questions dealt with in religion that are not already dealt with in secular science, philosophy, and ethics. The question of whether God exists, say, is a question of philosophy; we do not need to turn to religious traditions in order to discuss it fruitfully (similarly for the question of the meaning of life). In other words, secular rationality exhausts the range of humanly answerable questions.

4. What do you consider to be your own most important contribution(s) to theorizing about science and religion?

I would not describe my views about science and religion as "theorizing" at all—no *theory* is involved. I am simply making some rather obvious observations. What is perhaps unusual about my approach is that I don't think science is the real issue at all; the issue is about rationality in general, where this includes philosophy. There is nothing wrong with pursuing non-scientific questions, since not all questions are scientific. The trouble with religion is not that it fails to live up to scientific standards or uses a non-scientific method—though certainly it typically contains a good deal of obsolete science (as when it embraced the geocentric hypothesis). For philosophy itself isn't a type of empirical science either (I would call it a "formal science"). The trouble is that it fails to respect the rationality of science *and* philosophy: it sets itself up as immune to rational criticism, by reliance on notions like faith and divine revelation. Rejecting the dogmas of religion should not lead us into the dogma of scientism—the view that all worthwhile or meaningful questions belong to empirical science. Religion is not wrong to think there are questions beyond the range of science; it is wrong in its rejection of rationality as the only means of inquiry (as well as about specific matters, of course).

Nor should we suppose that "non-scientific" implies "supernatural": there is nothing supernatural about secular analytical philosophy, as many a disappointed sophomore can attest. When a student is asked to consider, for example, whether causality involves some kind of "necessary connection" (in a course on David Hume, say), he or she is not indulging in supernatural reveries; the question is as "natural" as can be. It is simply about the nature of causality. Speaking personally (and crudely), I don't like religion, but I don't like scientism either—both strike me as rife with error (and both are fundamentally irrational). Science, of course, is just fine; but scient*ism* is bad philosophical overreaching. In fact, it smacks of religion in its zealotry and imperviousness to impartial reason. I find scientistic zealots no more agreeable than religious zealots.

5. What are the most important open questions, problems, or challenges confronting the relationship between science and religion, and what are the prospects for progress?

I would like to divide my answer to this question into two parts. First, what is in the best interests of religion? The answer is obvious: stop trying to contradict established science. Opposing evolutionary science, say, is intellectually and politically disastrous. If religion is to survive, it must render itself consistent with scientific knowledge. So it must clearly jettison everything that conflicts with science. I don't think this is very difficult to do, and many sects have already made the necessary concessions and adjustments. You can certainly be a Christian and a follower of Darwin—you merely have to reject the Old Testament creation story.

Second, can religion co-exist with a proper respect for rationality? I don't believe it can, and therefore I think it should simply wither away. So long as religion relies on the idea of faith (revelation, scriptural authority) it is inconsistent with the demands of reason: but reason is sovereign. There cannot be reasonable beliefs without reason. The fundamental problem for religion is that there is no convincing reason to believe that God exists, so to believe in God is to make a "leap of faith." But to make a leap of faith is just to plunge into darkness—it is to abandon the guidance of reason. No amount of accommodation with science can change this central problem with religion. The belief in miracles illustrates the point perfectly: there is no reason to believe in miracles, and plenty of reason to doubt them, so anyone who insists on the reality of miracles is ipso facto unreasonable—they are being irrational. But being irrational is never defensible. Therefore the typical religious believer is indefensibly irrational. There is no way out of this elementary bind—no amount of hand waving about "faith" and "reve- lation" can gainsay it.

But what, you may ask, can be put in the place of religion, once the withering has occurred? Some may say that science is the right filler, but I think this is a bad idea. Science simply does not address itself to many of the proper concerns of human beings, particularly issues of va- lue and right and wrong. You are not going to get ethics out of physics and chemistry, or even biology and neuroscience. Science is wonderful and indispensable, but it is not enough. The putative choice between science and religion is a false choice, because there are many other are- as of legitimate human interest that fall under neither heading. We also have philosophy, literature, art, music, dance, gardening, tennis, stamp collecting, and a thousand other worthwhile activities. None of these is science, but neither do they require the irrationalities of religion. And many of them can serve at least some of the purposes of religion, pro- viding value, inspiration, meaning, and devotion. I myself would re- commend replacing religious education with philosophical education, starting in school. This would serve as a bulwark against the temptati- ons of irrationality and as an alternative to a desiccated scientism. Phi- losophy and religion inhabit roughly the same region of logical space, so secular philosophy can serve many of the purposes traditionally as- signed to religion—especially questions of value and meaning. It also gives expression to the human desire to *speculate*. Philosophy deals with the "big questions," as religion purports to.

I think that one of the biggest challenges facing us in the next century or so will be absorbing the withering of religion without falling into scientism. Growing numbers of people will find themselves out of sym- pathy with religion, and science will be perceived as the natural alterna- tive—no doubt aided and abetted by various kinds of propaganda, sub- tle and unsubtle. We already see signs of scientists trying to engineer takeovers of traditional areas of humanistic study—philosophy, ethics, and aesthetics. Many scientists today seem ignorant of philosophy and contemptuous of it; some even seem to equate it with religion (as with the subject called "metaphysics"). This strikes me as a dangerous trend. By all means, people should know their science; but that is not *all* they should know. Wouldn't it be terrible if, in a couple of centuries, human beings looked back at their religious past as an era of intellectual freedom, compared to the arid and limited scientism in which they are now daily indoctrinated?

15

Alister McGrath

Alister McGrath is Professor of Theology, Ministry and Education at King's College London, and head of its Centre for Theology, Religion, and Culture, and Senior Research Fellow at Harris Manchester College, Oxford. Until 2008, he was Professor of Historical Theology at Oxford University and in 2014 will take up the Andreas Idreos Professorship of Science and Religion at Oxford University. After taking First Class Honours in Chemistry at Oxford University in 1975, McGrath worked for several years in the Oxford laboratories of Professor G.K. Radda FRS, gaining a DPhil in 1978. He was then awarded First Class Honours in Theology at Oxford in 1978. McGrath was later awarded an Oxford Doctorate of Divinity (DD) in 2001 for research in the field of historical and systematic theology, and an Oxford Doctorate of Letters (DLitt) in 2013 for research in intellectual history, especially the relation of science and religion. He is the author of numerous books including *The Foundations of Dialogue in Science and Religion* (1998), *The Unknown God: Searching for Spiritual Fulfilment* (1999), *A Scientific Theology: Volumes 1-3* (2001-3), *Dawkins' God: Genes, Memes and the Meaning of Life* (2004), *The Twilight of Atheism: The Rise and Fall of Disbelief in the Modern World* (2004), *The Order of Things: Explorations in Scientific Theology* (2006), *The Dawkins Delusion?* *Atheist Fundamentalism and the Denial of the Divine* (2007), *The Open Secret: A New Vision for Natural Theology* (2008), *A Fine-Tuned Universe? The Quest for God in Science and Theology* (2009), *Heresy: A History of Defending the Faith* (2009), *Darwinism and the Divine: Evolutionary Thought and Natural Theology* (2011), *Surprised by Meaning: Science, Faith, and How We Make Sense of Things* (2011), and *Mere Apologetics: How to Help Seekers and Skeptics Find Faith* (2012).

---

1. What initially drew you to theorizing about science and religion?

I went up to Oxford University in 1971 to study chemistry, and subsequently did doctoral research in molecular biophysics under the supervision of Professor Sir George Radda. To begin with, I was an atheist, who took the view that science eliminated any credible basis for belief in God. However, I came to the view that this was a hasty judgement, not adequately grounded in evidence or argument. In part, this resulted from an immersion in the literature concerning the history and philosophy of science, which eroded the crude scientific positivism that I had uncritically absorbed from atheist writers, such as Bertrand Russell, and indicated that the question of evidential justification was much more complex than I had been led to believe. I came to the view that belief in God was more satisfying, existentially and cognitively.

I found this insight liberating, opening up new ways of thinking that were more imaginatively engaging and intellectually refreshing than the "glib and shallow rationalism" (I here borrow a phrase from C. S. Lewis) I had imbibed as a rebellious teenager. This meant a re-examination and recalibration of my views on science and religion, which had hitherto been rather simplistic—the standard "science disproves God" sort of thing.

I therefore decided to take my scientific education to doctoral level, and then move on to the detailed study of Christian theology. Only in this way, I had realized, would it be possible to theorize about science and religion in any meaningful and informed manner. Happily, a scholarship provided by Oxford University allowed me to take a first degree in theology. In the end, however, I was unable to secure funding for a doctorate in theology, and had to settle for higher Oxford doctorates awarded on the basis of publications.

I left Oxford in 1978 to go to Cambridge, where I had originally intended to do research on the Copernican debates. However, wiser advice prevailed, and I instead focussed on mastering the Christian intellectual tradition, especially in the early modern period, and the foundations of the shifting intellectual cultures of the Renaissance and early modern age. It was not until 1995 that I felt able to begin theorizing about the relation of science and religion with any intellectual credibility.

2. Do you think science and religion are compatible when it comes to understanding cosmology (the origin of the universe), biology (the origin of life and of the human species), ethics, and/or the human mind (minds, brains, souls, and free will)?

Science and religion can be forced into whatever mould that suits the needs and agendas of polemicists. I see no historical grounds for adopting an essentialist view of either science or religion which would be necessary if one were to make global judgements about their mutual relationships. The category of "religion," for example, is clearly non-empirical, depending on socially constructed and negotiated boundaries which ultimately reflect cultural and historical precommitments and prejudices. (This is one of the reasons why the "psychology of religion" encounters such difficulty, in that it tries to use empirical methods to study something that is not itself empirically disclosed and defined. Ultimately, it depends upon a shared—and contestable—definition of what "religion" might be.)

While I appreciate the utility of simplifications for polemical purposes, they cannot be sustained at a scholarly level—a point especially evident in the persistent repetition of the generalized "warfare of science and religion" myth, which lost intellectual credibility some time ago. There are points at which individual sciences and religious traditions find themselves in conflict; others at which they find themselves converging. There is no simple picture, despite the frequent media slogans suggesting that this is the case.

It is, I think, also important to appreciate the fluidity of the natural sciences, which are best thought of as a work in progress. It is perfectly reasonable for scientists to tell us what they believe the current thinking about various issues might be, and what evidence underpins those views. Yet science is on a journey. As further evidence accumulates, and theoretical development proceeds, the judgements of one generation may be overturned, and replaced with something quite different.

A century ago, for example, the dominant cosmological model was that of an essentially static universe. Religious language about "creation" made no sense. That consensus enabled Bertrand Russell to defeat his Christian opponent, Frederick Coplestone, in the great BBC debate of 1948 about God. Although Coplestone won the debate about whether God was required to safeguard moral values, Russell gained an easy victory in the debate about whether one needed to invoke God to explain the universe. All that, of course, has changed radically since the Second World War, as evidence accumulated that the universe had an origin. The debate is now whether origination of the universe entails its creation. It is an important development, in the light of some poorly informed assertions that the advance of science necessarily entails the erosion of the plausibility of belief in God.

In other areas, tensions remain. Conservative Muslims and Protestant Christians reject the idea of biological evolution, generally because this is held to contradict their sacred texts. I cannot comment on Islam. In the case of Christianity, the Bible is open to interpretations sympathetic to biological evolution, which have a long history of use in the Christian tradition—such as the approach developed by Augustine in the year 401, which has gained a new following in recent years. This ancient way of reading the biblical creation accounts contrasts with the more literal interpretations that arose within Protestantism since about 1700, indicating that the real debate concerns issues of biblical interpretation, not the idea of evolution itself.

I fail to see that science *qua* science has anything important to say about ethics. Despite its intentions, Sam Harris's recent book *The Moral Landscape* ends up confirming this judgement, affirming some platitudes of scientific ethical benevolence without actually justifying them. I find myself in complete agreement with Richard Dawkins (and many others) who make the entirely reasonable point that science is unable to tell us what is right or wrong. That's not its role, and it loses its way if it tries to. Good science steers clear of moral debate; its role is better seen as illuminating and informing a wider cultural debate.

Despite being dogged by over-interpretation of empirical evidence, recent discussions about the mind, consciousness, and free will have the potential to prove both interesting and productive. But it's too soon to say what the outcomes will be. We must wait for evidence to accumulate, and mature judgements to emerge.

3. Some theorists maintain that science and religion occupy non-overlapping magisteria—i.e., that science and religion each have a legitimate magisterium, or domain of teaching authority, and these two domains do not overlap. Do you agree?

No. The idea that science and religion do not overlap is simply wrong, and demonstrably so. It's a neat little device, ideally adapted to the political needs of the American academy, which serves the needs of those who want to keep religion out of science, and science out of religion. But things are more complicated than this. It is perfectly reasonable to argue that, at least in certain respects, science *ought* not to interfere in religion, and religion *ought* not to get involved in scientific debates. But it is a simple matter of fact that there is traffic between the two domains, whether we think this is a good thing or not. If we were to think in terms of a Venn diagram, I suspect that the extent of overlap between science and religion might be less than some might think. But it's there.

What seems to me to be important is to develop a theoretical framework which both acknowledges this reality, and tries to place it on a more rigorous footing. My own research has focussed on the importance of "critical realism," as developed by social scientists such as Roy Bhaskar to deal with the complexity of reality. This approach recognizes that reality is stratified, and that different levels of explanation are called for within each domain. We live in a complex, multi-layered universe. Each level has to be included in our analysis. Physics, chemistry, biology and psychology—to note only four sciences—engage with different levels of reality, and offer explanations appropriate to that level. But they are not individually exhaustive. A full picture results when these different levels of explanation are woven together.

A comprehensive explanation must bring together these different levels of explanation, in that (to give an obvious example) the physical explanation of an electron is not in competition with its chemical counterpart. My Oxford colleague John Lennox, who is a mathematician and philosopher of science, uses a neat illustration to make this point. Imagine a cake being subjected to scientific analysis, leading to an exhaustive discussion of its chemical composition, and the physical forces which hold it together. Does this tell us that the cake was baked to celebrate a birthday? And is this inconsistent with the scientific analysis? Of course not.

A scientific description of the world describes how it arose from an initial cosmological event (the fiery singularity of the "big bang"), which led, over a long period of time, to the formation of stars and planets, creating conditions favourable to the origination and evolution of living creatures. No reference is made, or needs to be made, to God. That's one level of explanation. A Christian might want to speak of God bringing the world into existence, and directing it towards its intended outcomes. That's another level of explanation. They are not in competition; if anything, they are supplementary.

My atheist colleagues will, I imagine, wish to suggest that the second is unnecessary. But that's not my point. The main thing to appreciate is that the somewhat dogmatic assumption that science and religion offer *competing* explanations needs to be challenged. The possibility of mutually enriching explanations, operating at different levels, must be taken seriously. For some, there cannot be multiple explanations of the same things, and only the scientific explanation can be valid. But there are other ways of looking at this question.

Some, of course, would say that religion explains nothing—I think, for example, of the late Christopher Hitchens, who had strong views on this matter. However, the rhetorical force of his bold assertion is somewhat reduced by his failure to explain what is to be understood by "explain." One of the most interesting points made by William Whewell, one of the most important nineteenth century philosophers of the empirical sciences, concerns the capacity of a good theory to "colligate" observations, like a string holding together a group of pearls in a necklace. Whewell held that all observation involves what he terms "unconscious inference," in that what is observed is actually unconsciously or automatically interpreted in terms of a set of ideas. A good theory is able to "colligate" observations that might hitherto have been regarded as disconnected.

We might think, for example, of Newton's theory of gravity as "colligating" observations that had up to that point been seen as unconnected—such as the falling of an apple to the ground, and the orbiting of planets around the sun. As many of you will know, this idea of explanation as providing a "big picture" which permits the colligation of what might otherwise be seen as unrelated and disparate events underlies Margaret Morrison's recent notion of unitative explanation. It also underlies the approach known as "inference to the best explanation," now widely regarded as the dominant philosophy of science.

For me, Christianity offers a "big picture" of reality which is intellectually hospitable towards the natural sciences. It offers a colligative framework that allows things to be framed within a wide context. That's what drew C. S. Lewis to believe in God, after an extended time as an atheist. As he remarked at the time of the root causes of this transition:

"I am an empirical theist. I arrived at God by induction."

4. What do you consider to be your own most important contribution(s) to theorizing about science and religion?

I would like to think that my most important contribution to the science and religion debate is to try and rebut the overstatements and misunderstandings which prevent the discussion from getting anywhere interesting. There is nothing more wearying than the gross polemicized oversimplifications and (I trust, accidental) misrepresentations which litter and hinder what has the potential to be a positive and critical discussion.

It is utterly depressing, and pedagogically demoralizing, to watch scientifically illiterate theologians and religiously illiterate scientists squabbling pointlessly over misunderstood ideas relating to their interface—especially when that misunderstanding arises from a contemptuous refusal to understand alternative approaches, or to ridicule them and those who adopt them. Happily, things are now improving, and a younger generation is emerging which is willing to try and understand other perspectives, rightly realizing that understanding does not entail agreement. It's long overdue, and I am hopeful about the future direction of such discussions.

When not waving my "It's more complicated than this!" banner around, my main contributions concern exploring the implications of recent developments in the philosophy of science, and developing the traditional notion of "natural theology."

5. What are the most important open questions, problems, or challenges confronting the relationship between science and religion, and what are the prospects for progress?

All the big questions are still open for discussion. Are we predisposed to believe in God, as Christian theology has always asserted? What is the impact of our evolutionary past on our present intellectual and moral habits and intuitions? How may we define "religion"? Is metaphysics compatible with empiricism? Is it meaningful to speak of "purpose" in life, or within the cosmos in general? The big questions may ultimately prove to be unanswerable, to everyone's satisfaction. But the process of engagement and reflection is not merely enjoyable, but also illuminating and enriching.

One of the key components of any future dialogue between science and religion is a recognition and respect of limits. Religion isn't science; science isn't religion. Although there are conspicuous and noisy exceptions, on the whole scientists are much better at acknowledging the limits of their discipline than either theologians or philosophers of religion. I'm a great admirer of the writings of the biologist Peter Medawar, and welcome his emphasis on the primacy of the empirical within the scientific method, with a corresponding recognition of the limits that this places on the method. For this reason, Medawar argues—surely rightly—that what Karl Popper termed "ultimate questions" (e.g., "why are we here?") remain, and will remain, beyond the scope of the scientific method. I have always been impressed by this intellectual humility, which is not merely intellectually appropriate, but is hospitable towards dialogue and discussion.

The prospects for progress in such discussions about the big questions of life are excellent, once we can get an intelligent conversation under way. By this, I do not mean that we will necessarily reach agreement, but that we may be able at least to gain a better understanding of the issues. I'm always amused by a quote from Robert Jastrow's book *God and the Astronomers*, published around the time I was getting interested in the interaction of science and religion. There's enough truth in it to allow it to stand as the conclusion to my reflections in this piece.

> Science will never be able to raise the curtain on the mystery of creation. For the scientist who has lived by his faith in the power of reason, the story ends like a bad dream. He has scaled the mountains of ignorance; he is about to conquer the highest peaks; as he pulls himself over the final rock, he is greeted by a band of theologians who have been sitting there for centuries.

16

Mary Midgley

Mary Midgley is an English moral philosopher known for her work on science, ethics, and animal rights. She retired as a Senior Lecturer in Philosophy at Newcastle University in 1980. She has written extensively about what philosophers can learn from nature, particularly animals. She has also written in favour of a moral interpretation of the Gaia hypothesis. Midgley strongly opposes reductionism and scientism, and *The Guardian* has described her as the UK's "foremost scourge of 'scientific pretension'" (2001). Her books include *Beast and Man* (1978), *Animals and Why They Matter* (1983), *Wickedness* (1984), *The Ethical Primate* (1994), *Evolution as Religion* (1985), *Science as Salvation* (1992), and *The Owl of Minerva* (2005).

1. What initially drew you to theorizing about science and religion?

I don't theorize about them in the sense of formulating and opposing theories. I watch their interactions with a concerned interest and try to work out suggestions about them when I think they are needed.

My philosophical aim has always been to find a sensible middle path between extremes which are driving each other to unworkable positions. In my first book, *Beast and Man*, I tried to mediate between a rather coarse, scientist view of human nature (Desmond Morris, later succeeded by Richard Dawkins) and an unrealistic Existentialist exaggeration of human freedom (J-P Sartre) because, in the 1960s, these were the main doctrines contending for the public's loyalty and they expressed clashing views about that central topic, the meaning of Freedom. I wanted to point out the elements of truth in both, rather than continuing the older debate between evolutionism and religion, of which I was, of course, aware but which did not then seem to me to be getting anywhere.

I still think that, today, forms of these two doctrines provide the main arena of conflict in people's minds, even though different prophets are now active there. In particular, the supposed demise of Existentialism has not extinguished the kind of bogus individualistic spirituality that was associated with it. This now centres instead on various forms of egoism, producing a mystique of selfishness (e.g. Ayn Rand and Richard Dawkins, who wove this mystique of selfishness deep into the fabric of *The Selfish Gene*).

I do not have any special interest in the conflict between science and religion as such. I am interested in each of them separately but their functions seem to me quite different. The conflict is, I think, a by-product of various political clashes which have resulted from the churches becoming centres of social power and have led people to think of them as rivals for a single slot. The 'new atheists' strike me as simply repeating the well-known themes of this unreal warfare, sometimes reciting old political objections (often justifiable ones) to this church-centred power and sometimes just exhibiting a tin-eared ignorance about the motives that lead people to religion.

2. Do you think science and religion are compatible when it comes to understanding cosmology (the origin of the universe), biology (the origin of life and of the human species), ethics, and/or the human mind (minds, brains, souls, and free will)?

It depends on what you mean by religion. Buddhism does not, as far as I know, clash with any reputable cosmological, biological or neurological doctrine. Nor, I think, does Hinduism. They both do, of course, say things about the nature and destiny of souls which form no part of these various sciences, but this is not a clash. It is a quite proper separation of provinces.

By contrast, fundamentalist Christianity and Islam, which both require a literal interpretation of the Bible, do clash with any cosmological doctrine that is scientifically credible today. This is not surprising seeing that they were formulated long before modern science was invented, but it is a sufficient reason for not accepting them now.

If, however, one grasps the symbolic intent of the Biblical writers and understands Christianity (or Islam) as an incomplete, perpetually growing attempt to grasp God's plan, this sort of clash probably need not arise. What is certain is that God cannot be calling on us to accept lies. If solid evidence for the past development of the world indicates that it occurred gradually over many aeons—which it apparently does—then we have to conclude that this is true and that the account given in Genesis has a spiritual, not a physical meaning. Creation, in fact, was a rather more complicated and longer business than our forefathers supposed, and this is not very surprising.

As for supposedly scientific accounts of the mind, they are seldom relevant to it because the physical sciences can deal only with physical questions and minds are not physical items. Indeed they are not *items* at all; they are sets of activities that are essential for people's lives. Biologists can, of course, speculate, as anybody else can, about the meaning of our experiences. But they can't use the reductive stance of claiming that their physical discoveries provide a complete account of experience and thought. Those discoveries are plainly answers to questions of a special kind, not general explanations.

For instance, current theories about the behaviour of neurones do indeed tell us interesting things about the apparatus which makes thought physically possible, and this can often be important, especially when the apparatus goes wrong. But the thoughts themselves can only be understood by considering their connection with other thoughts and so with the world that we think about—that is, by more thinking. Those scientists who—along with a few philosophers—claim to have proved that the inner life, or the self, is an illusion and neurones are the reality behind it need to think a little harder about what question it is that these suggestions are supposed to answer.

3. Some theorists maintain that science and religion occupy non-overlapping magisteria—i.e., that science and religion each have a legitimate magisterium, or domain of teaching authority, and these two domains do not overlap. Do you agree?

Talk of magisteria and teaching authority may seem to locate this issue as a power-struggle among academics. What is needed, however, is guidance for the individual soul that wants to know what it should believe. This it has ultimately to decide for itself.

Stephen Jay Gould proposed this language of magisteria in a laudable attempt to stop religious authorities interfering with science. It expresses the solid general point that prophets should not be believed when they talk about matters beyond their own expertise. But the question whether certain particular religious and scientific doctrines can conflict needs to be resolved separately by each person in each particular case. It can't rest on the classification of different disciplines that reigns in universities.

If I'm asked what I do myself, I consider the kind of evidence that there is on both sides of the question and compare the two. If it seems that I need to change my general attitude, I do so. There is no general recipe.

4. What do you consider to be your own most important contribution(s) to theorizing about science and religion?

I think my own modest contribution to this discussion probably centres round my pointing out that Science itself is not a Religion—that the sort of mindless acceptance which may have been associated with religion does not become proper or laudable when it is transferred to what is mistakenly believed to be science. I have made this point most notably in two books, *Evolution as a Religion: Strange Hopes and Stranger Fears* and *Science as Salvation: A Modern Myth and Its Meaning*.

The kind of scientistic fantasizing that I tried to expose in these books comes, I think, from an excessive faith that something vaguely called 'science' can get rid of all the rest of thought by simplistic reductive methods. This is something I have consistently opposed because I think it is a chronic evil of our age. 'Reduction' of one study to another is quite proper when it merely involves translating the propositions of one into the language of another that is closely related to it, as in the reduction of chemistry to physics. When it is attempted over wider gaps, between more diverse studies, it becomes a sort of metaphor which may well have no effective application at all. The real successes of physical science have unfortunately led people now to suppose readily that it has the answer to everything, and so to accept silly opinions from scientists on non-physical questions, such as the existence of the self.

5. What are the most important open questions, problems, or challenges confronting the relationship between science and religion, and what are the prospects for progress?

I think that this absurd over-confidence in science and technology—which are treated as synonymous, and are believed to make religion unnecessary—is a fearful weakness in our civilization and may well lead to its downfall. Its folly should be exposed by all the areas of thought that interact with it, including science and religion.

Even though the various religions in which people have so far put their trust all have their faults, they do at least direct public attention to serious moral and social questions. Those questions are now widely believed to have been solved by something vaguely imagined as science.

We can see how little this over-confidence has to do with real science when we notice the astonishing phenomenon of climate-change denial. Here the solid and alarming verdict of serious scientists is brushed aside by people who simply refuse to believe that our civilization could be in any real danger. This self-glorification does appear to be a kind of religion; a habit of worshipping ourselves and our culture. It travesties the name of science by applying it to belief which are radically anti-scientific. I can only hope that people are beginning to see through it.

17

Seyyed Hossein Nasr

Seyyed Hossein Nasr, currently University Professor of Islamic Studies at The George Washington University, is one of the most important and foremost scholars of Islamic, Religious, and Comparative Studies in the world today. Author of over fifty books and five hundred articles—which have been translated into several major Islamic, European and Asian languages—Professor Nasr is a well-known and highly respected intellectual figure both in the West and the Islamic world. He was the first Muslim to deliver the prestigious Gifford Lectures, and in 2000 a volume was devoted to him in the Library of Living Philosophers. His books include *Islam and the Plight of Modern Man* (1976), *Islamic Science* (1987), *Knowledge and the Sacred* (1989), *Traditional Islam in the Modern World* (1990), *Science and Civilization in Islam* (1992), *The Need for a Sacred Science* (1993), *An Introduction to Islamic Cosmological Doctrines* (1993), *Islam: Religion, History, and Civilizations* (2002), *Islamic Philosophy from Its Origin to the Present* (2006), *Man and Nature: The Spiritual Crisis in Modern Man* (2007), *Islam, Science, Muslims, and Technology: Seyyed Hossein Nasr in Conversation with Muzaffar Iqbal* (2009), and *Islam in the Modern World* (2011).

---

1. What initially drew you to theorizing about science and religion?

I was born and brought up in a Muslim family in Persia where the presence of religion, in my case Islam and not Christianity or some other religion, was very strong. At the same time my father was a famous physician in addition to being a renowned scholar, philosopher and ethicist, who also knew French well and was very familiar with modern science. In our house, in addition to such Persian intellectual and cultural heroes as Ibn Sīnā, who was at once a scientist and a philosopher, and the Sufi poet Rumi, such European scientists as Descartes, Pascal and Berthélot, who were also men devoted to religion, were made known to me by my father. As a young boy I felt no tension between religion, that is, Islam in which my life was immersed, and the modern Western sciences which my father was encouraging me to learn and which we were studying at school. Nor did the religion and science issue arise when I came to the Peddie School in America to complete my secondary education. I did very well in both scientific subjects and the humanities in a school which was Baptist and emphasized religion. I even had to go to church every Sunday, although I was a Muslim and I prayed on my own even when in the church, in a house dedicated to Christ. My faith in God and love for religion continued during those years when I was outwardly separated from my own religious community.

It was at M.I.T. to which I went to study physics and mathematics after graduation from Peddie that a crisis began in my mind concerning the philosophical conflict between modern Western science and religion. I was too philosophically minded to simply leave the issue alone. As the result of this intellectual and spiritual crisis, I began to take more courses related to this issue and also to read everything on this subject that I could get my hands on from the works of Bertrand Russell and A. N. Whitehead to those of such Catholic writers as Jacques Maritain and historians of science and philosophy as Alexandre Koyré, with the last two of whom I took courses at M.I.T. and Harvard. My search also went beyond the religion and science issue only in the Western context and embraced non-Christian religions and also non-Western cosmologies and natural sciences, especially the Islamic. Since then, that is during over the past half century, the religion and science question has been at the forefront of my intellectual concerns.

2. Do you think science and religion are compatible when it comes to understanding cosmology (the origin of the universe), biology (the origin of life and of the human species), ethics, and/or the human mind (minds, brains, souls, and free will)?

No, I do not think that they are compatible if by science we mean modern science which is limited to only the quantitative level of existence and excludes higher levels of being, and by religion we mean integral religion which includes of necessity teachings about cosmogenesis, cosmogony, cosmology, anthropology and the world of nature and not only ethics and eschatology. Modern scientific cosmologies seek to explain everything in terms of a truncated view of reality and are based often on conjectural extrapolations. That is why every few years a new scientific cosmological scheme appears upon the scene without this being a real "progress" in cosmological ideas, very different from what we see let us say in chemistry. As for the origin of life and of the species, again "scientific explanations" are based on the false presumption that there is only one level of reality in terms of which alone must these issues pertaining to life be explained. When it comes to the question of evolution, even some eminent Western philosophers of science such as Karl Popper have stated that this theory does not meet the conditions of scientific verifiability. And yet, no scientific theory has done as much as evolutionary theory to destroy the religious teachings about the origin of life and of various species even though there is not a single paleontological evidence of macroevolution.

When we come to ethics, modern science, if understood correctly, is neutral in this matter while ethics is central to religion and all the ethical teachings of various civilizations from the Chinese and Japanese to the

Western have a religious foundation. In this context it is only fair to state that many scientists are personally very ethical people, but then there are other scientists who make weapons which obliterate the life of numerous innocent people in the blink of an eye. There are also, of course, those who use religion to condone unethical actions. But this fact itself can only be explained by the religious understanding of human nature and the reality of evil in human life.

When it comes to the question of the mind and consciousness, there are those who are trying to study it "scientifically," but the scientific *Weltanschauung* does not really have a clue as to what consciousness is and so takes recourse to one form or another of reductionism. Furthermore, those few *bona fide* scientists such as Rupert Sheldrake who try to step outside this reductionist paradigm are attacked and ostracized by the "scientific community" with its own self-proclaimed "scientific orthodoxy." In contrast religion sees consciousness as an attribute of the spirit that belongs to a realm of reality other than what modern science claims to be real.

3. Some theorists maintain that science and religion occupy non-overlapping magisteria—i.e., that science and religion each have a legitimate magisterium, or domain of teaching authority, and these two domains do not overlap. Do you agree?

Theoretically that might be a possibility if modern science were to accept that it is in fact *a* and not *the* science of the world of nature, but in practice this does not happen. Modern science, as perceived and interpreted by most if not all of its practitioners, is monopolistic, triumphalistic and totalitarian and claims for itself the exclusive knowledge of nature and by extension of "reality," relegating the religious view of nature to the realm of sentimentality, art, poetry, etc., but not science. Moreover, if we look at the history of the relation between religion and science in the West since the 17th century, we see that they have not stood as two contending armies staring at each other in the field. Rather, modern science has continued its onslaught into the religious universe taken in its totality, desacralizing one domain of reality after another from the firmaments and the corporeal world around us to life and the human body and more recently to the psyche and the mind according to a worldview and methodology in which the category of the sacred is strictly speaking meaningless. For most of its practitioners to explain is to explain away the mystery of existence (although there are some notable exceptions), to remove the sense of wonder, to substitute the understanding of the mechanism by which something functions for the total reality of that thing, all of this done through reductionism. If a complete metaphysical and cosmological knowledge and a total philosophy of nature were to become once again foundational in today's world, then modern science could be integrated into a hierarchy of knowledge in which the magisteria of both modern science and religion could be accepted without confrontation and mutual rejection. But such a situation is not possible within the paradigm that reigns over modern thought today.

4. What do you consider to be your own most important contribution(s) to theorizing about science and religion?

My most humble contributions in this domain have been the following: (a) Reviving traditional metaphysics and cosmology and in their light revealing the shortcomings of the totalitarian claims of modern science and also trying to show what modern science is and what most of its practitioners and popularizers claim it to be but it is not. (b) To bring back to life other philosophies of nature which stand outside the modern paradigm and are neglected or rejected by it. (c) To extend the science and religion debate beyond modern science and Western Christianity by dealing with both non-modern sciences and other religions especially Islam. (d) To deal with this issue extensively in the Islamic context and seek to show how Islam can integrate modern science into its intellectual tradition without betraying the authenticity of its own worldview.

5. What are the most important open questions, problems, or challenges confronting the relationship between science and religion, and what are the prospects for progress?

The most important question and problems relate to the issue of the genesis of the cosmos itself, of life, of consciousness and of the laws that govern all domains of cosmic existence and of the approaches of religion to them on the one hand and the views of science on the other hand to these basic questions. The greatest challenge is to rediscover and accept a universal metaphysical paradigm within which both religion and science would have legitimate existence and could co-exist without claims of triumphalism and exclusivism.

With the current triumphant and global spread of modern technology which is identified by the masses with modern science, it is difficult to hope for "progress" in this domain. But there is still some hope and it comes from the present day environmental crisis. The unprecedented destruction of the harmony of nature through the application of modern technology, which is after all the application of modern science, has already caused many thinking people, especially in the West where this technology began, to try to look beyond the dominant paradigm of the natural sciences and to rediscover a holistic view of nature which of necessity includes its religious and spiritual dimensions. At the present moment I think that this crisis offers the best and perhaps the only hope for return to a sane worldview within which an in-depth harmony could be established between religion and science. I believe that anything less than the rediscovery of true metaphysics and cosmologies that embrace all levels of cosmic reality and are not bound to only the material and quantitative level of existence, results only in continued confrontation or at best a so-called harmony that is merely cosmetic and avoids the deeper issues involved.

18

Timothy O'Connor

Timothy O'Connor is an American philosopher well known for his work in metaphysics, philosophy of mind, and philosophy of religion. He is currently Professor of Philosophy at Indiana University Bloomington and has written extensively on free will, emergence, and philosophical theology. He is the author of *Persons and Causes: The Metaphysics of Free Will* (2000) and *Theism and Ultimate Explanation: The Necessary Shape of Contingency* (2008), the editor of *Agents, Causes, and Events: Essays on Indeterminism and Free Will* (1995), and the co-editor of *Philosophy of Mind: Contemporary Readings* (with David Robbi) (2003), *A Companion to the Philosophy of Action* (with Constantine Sandis) (2010), *Emergence in Science and Philosophy* (with Antonella Corradini) (2010), *Downward Causation and the Neurobiology of Free Will* (with Nancey Murphy and George Ellis) (2010), and *Religious Faith and Intellectual Virtue* (with Laura Callahan) (2014). He is currently working on a book on the integration of Christian faith, reason, and science.

"SCIENCE AND CHRISTIAN FAITH: THE PURSUIT OF AN INTEGRATED AND COMPLETE UNDERSTANDING OF HUMANITY AND THE COSMOS"

1. What initially drew you to theorizing about science and religion?

Any good philosopher whose principal interests lie in metaphysics and epistemology, as mine do, will have thought hard about the sorts of explanations that mature science does (and doesn't) provide, and what metaphysical commitments do (or don't) underlie them. She will also have thought about the metaphysics of theism (if only because so many important historical figures make theism central to their metaphysics and sometimes their epistemology) and about whether or not theism offers plausible forms of explanation of very general truths about reality. Finally, if she is herself a religious theist, as I am—I embraced the Christian faith at the same time that I began to study philosophy at university—she will think about the relationship of scientific and theistic explanations, not merely as concerns general aspects of reality, but also religious experience, purported miracles, and the like. That is, she will find herself deep in the thickets of 'theorizing about science and religion.'

A secondary motivation for engaging in this study is dismay at the state of public discussion of these matters. I am not *so* much surprised at the extent of fear or suspicion of sciences that touch upon human origins and human behavior among contemporary Christians (who run the intellectual and educational gamut), though I find it lamentable and, for those of any decent education, foolish. I am more surprised at the fairly common belief among secular *scholars* that there is a deep and unresolvable conflict between taking seriously both science and traditional religious belief—and I am especially surprised when this is the attitude of my fellow philosophers, who have the training to enable them to know better. I read very carefully reasoned criticisms of religious belief currently on offer; that is simply what philosophers are supposed to do and, in any case, I have no interest in constructing my life around a supposedly comforting lie. If there is good reason to think that Christianity, or mere theism, is false, I want to know it. There are some challenging criticisms of such commitments given by philosophers I respect. There are also some embarrassingly bad criticisms given wide public play, mostly by scientists who lack the ability to see where science leaves off and scientifically-informed philosophical reflection on reality begins, but also in some cases by philosophers. (I recently attended an exchange on the compatibility of science and religion by two prominent philosophers, since published as a book. Neither side made a particularly compelling case, but as the philosopher arguing the 'con' side spoke, I couldn't help but marvel at the high ratio of mere rhetoric to argument. I was sorely tempted to stand up and offer to finish the job, confident that I could argue his side a whole lot more cogently than he was doing!) It appears that many scientifically-motivated critics of religion don't bother to read the most careful proponents of reasoned faith. If I am correct in this judgment, it says nothing about the relative merits of religious and non-religious conceptions of reality. It merely reflects the common tendency to get sloppy when arguing a thesis that is already widely held by one's immediate peer group or target audience. Criticisms of the state of public play aside, I am committed to pursuing an integrated system of personal belief. For me, that means thinking hard about science and religion.

2. Do you think science and religion are compatible when it comes to understanding cosmology (the origin of the universe), biology (the origin of life and of the human species), ethics, and/or the human mind (minds, brains, souls, and free will)?

First, a brief comment on how I will be understanding, for present purposes, each of the three crucial terms: 'science,' 'religion,' and 'compatible.'

'Religion' covers an enormously varied range of human practices and beliefs (among other attitudes), and it is impossible to give the question, taken in its full generality, a straightforward answer. I will take it as shorthand for theistic religion, and more specifically Christian religion anchored in the Nicene Creed, just to give it a clear target that captures much of what is in view when one speaks of the compatibility of the sciences and Christianity. What I say about the question so construed will carry over to other forms of revealed monotheistic religion. I am concerned specifically with the beliefs and other attitudes that are *common* to Christian subtraditions that affirm the Nicene Creed. It is evident that some varieties of Christian belief are incompatible with modern cosmology and biology—those varieties that insist, on religious grounds, on the truth of young earth creationism, for example.

As for 'science,' we may take it to encompass not just the deliverances of various mature sciences (such as scientific theories, laws, and the identification of causal mechanisms), but also a commitment to the soundness of its characteristic methods of empirical investigation. Some who are skeptical of (some of) the deliverances of established scientific theories (e.g., natural evolution) profess adherence to what they deem the methods of science, claiming that 'mainstream' science gets things wrong by allowing underlying metaphysical commitments such as metaphysical naturalism to skew their interpretation of the evidence. I believe that this criticism is not well-founded. But even if they were right, they would not be thereby demonstrating the compatibility of their 'creationist' views with science as I am using that term, as the latter encompasses both characteristic methods and well-established results (there is vagueness on the boundaries of 'well-established,' but nothing of interest here hangs on that).

Finally, the question we are interested in requires us to give a broad, somewhat vague meaning to the word 'compatible.' The truth of a great many unreasonable ideas is *logically* compatible with acceptance of the results of modern science and its methods for investigating the world: no contradiction is generated by adding such ideas to the body of scientific commitments (assuming, nontrivially, that those commitments themselves do not entail contradictions). But compatibility in that sense is of little interest. What is interesting and much disputed is whether the commitments of creedal Christianity—that there is an omnipotent, omniscient, and wholly morally good supernatural being who is the source of everything else, our universe included, who has special regard for his human creatures, and who has miraculously acted in human history at various times and places, most spectacularly in his incarnation in Jesus of Nazareth on a rescue and restore mission—are *consonant* with what science teaches us. Put negatively, what I mean is that Christian commitments are not rendered highly unlikely by those teachings; put positively, one who grasps both Christian and scientific commitments might rationally affirm them jointly. We could spend a lot of time trying to pin down the operative notion of 'rationality' here, but I will take it as reasonably well understood.

Cosmology

Modern cosmology indicates that our universe is approximately 13.8 billion years old, expanding continually from a highly compressed state under the impetus of a small number of fundamental forces. *Perhaps* it is finite in measure but lacks a first moment. *Perhaps* it is a product (inevitable or evitable, as the case may be) of a prior physical reality— and perhaps one of a very large number of such products. And there are other scenarios besides that are open on current evidence and have attraction to at least some respected theorists. Is the well-established core, supplemented by any of the speculative 'additions' to it now in currency, consonant with creedal Christian belief, e.g., that God is the creator and sustainer of 'heaven and earth'?

I think the answer is evidently yes, and that the reasons for thinking so were articulated well long ago by Aquinas and Leibniz. Fundamental physics aims to give a unified, parsimonious, and accurate account of the basic structure and dynamics of physical reality. It has not fully realized its aim yet. But we can readily imagine in broad outline explanatorily ideal limit-case scenarios: On one, final physical theory bottom out in a single fundamental property and property-bearer, with a single, relatively simple and elegant equation governing the co-evolution of instances of the fundamental entity type through a beginningless space-time. On another, physical theory tacks in rather a different direction. Rather than burrowing down to simple foundations of a single universe, it spreads out. Satisfyingly complete explanation may be achieved, it is claimed, through the devising of an elegant and empirically adequate theory that locates our universe within a vast structure of totalities that exhibits completely non-arbitrary properties: a plenum of (largely) disjoint island universes. This way's limit case involves the existence of all mathematically consistent totalities: all possible universes, as MIT physicist (and closet metaphysician) Max Tegmark (2008) proposes. Verifying either of these limit scenarios would be a spectacular explanatory achievement. But neither would yield a *complete* explanation of physical reality, one that involves no brute givens, leaves no explanatory loose ends whatsoever. For the most fundamental contingent fact that physical reality *is as described* would necessarily be left unexplained. Even a plenitudinous multiverse invites the question: why is there this multiverse? Why not just one universe, or seventeen, or none at all? (These questions are *not* to be understood as requesting purposive explanation—just explanation, in any of its basic varieties.) To answer this sort of question, we must pass from physics to metaphysics. Classical theism proposes the schematic answer: God, a necessary being, and the fount of all possible contingent reality, freely and purposively willed into being that which is, having been capable of willing into being any of a variety of alternative possibilities. If an explanation conforming to that schema were true, God explains that which fundamental science does not—and cannot. Empirical science provides metaphysically contingent, not necessary explanations, and a bit of reflection shows that an adequate explanation of contingent existence must come from outside the connected web of contingent existences, from a realm of necessity. Modern cosmology, then, whatever its future course, can pose no threat to Christianity's claim that God is the ultimate source and sustainer of the reality we inhabit. To be sure, there are alternative *metaphysical* explanations one might propose, including questioning the claim that our universe (or multiverse) exists only contingently. But these challenges to theism come from alternative metaphysics, not physics. (For a view on the relative merits of various metaphysical explanations of ostensibly contingent reality, see my *Theism and Ultimate Explanation*.)

Biology

I pass on to modern biology. It has been well established that all forms of life now present on earth are linked through a common ancestry. From life's origin on earth some 3.5 billion years ago (which is not yet well understood), all subsequent living organisms descended through endless cycles of reproduction and maturation involving small modifications in genetic information, some of which took hold and led to stable variants in a way that natural selection describes, eventually leading to a vast tangled tree of life. We modern humans occupy one offshoot of that tree, which is more like a sprawling bush than a Christmas tree with an apex. We are the product of a great many adaptations in ancestral hominid and pre-hominid species to the contingent challenges and opportunities of specific environments—better, one outcome of a co-evolution of diverse life forms inhabiting the same environmental niches. We have a great many finely honed biological and cognitive assets alongside various biological liabilities, the result of inevitable trade-offs in the natural selection process. Our adaptation as a species slowly continues. Life itself is now understood to be an extraordinarily complex biochemical process, an interplay of myriad cellular and intercellular mechanisms with 'top-down' organismic control parameters as emphasized in the recent resurgence of systems biology.

This revolution in our understanding of human life and its place in the natural world poses several prima facie challenges to traditional Christian theology. I have space to consider but three, and these only briefly. First, it is sometimes thought that the non-purposive character of biological evolution and its central mechanism of natural selection contradicts the theological claim that God purposively created us. This thought, while often expressed by educated, thoughtful minds, is easily seen to be mistaken. Evolution by a slow process of descent with modification is contrary to the common picture of God's 'specially' creating humans in a virtually instantaneous, supernatural manner. But that picture, based of course on a literal reading of the creation narratives in the first two chapters of *Genesis*, is not a creedal commitment, and it wasn't even held by all Christian theologians before Darwin. (No less significant a figure than St. Augustine [354–430] did not hold it.) Furthermore, modern biblical scholarship gives us very good reason to think that such a literalist construal of *Genesis* rests on a misunderstanding of the genre from which it arose. In any case, Christians (and theists of every stripe) can maintain that God's purposive creation of human beings is manifested in his creating and sustaining a world with fundamental laws that undergird the very biological processes that in the fullness of time gave rise to us. Natural evolution is not intrinsically purposive, but that is fully consistent with its being extrinsically purposive—caused to unfold in accordance with the intentions of the designer and creator of the whole universe, not just the biosphere.

Some say that a more careful consideration of natural history offers a less easily resolvable challenge to the thesis that God's providentially designing our universe's basic physics suffices for his purposively bring about *us*, specifically. Evolutionary developments, they say, are fraught with contingencies, such that things could have easily gone in very different directions, with very different kinds of creaturely outcomes. Evolutionary fitness is relative to one's environmental niche, and small changes in the actual environments faced by myriad species would have led to very different selection pressures and so outcomes. To take one dramatic example, if the meteor that apparently struck our planet sixty-five million years ago had instead been a near-miss, the planetary climate would not have undergone the radical change it did (becoming much colder), making vegetation in various regions relatively scarce. The near-miss scenario could easily have happened. But the non-inevitable meteor strike did occur, and the ecological changes that it brought about were important factors in making small mammals to be comparatively better adapted to their environments than large dinosaurs, so that the dinosaurs waned and mammalian life flourished. We, in short, were not inevitable. If God did intend us to be among the outcomes of a long evolutionary process, He got lucky, as it was unforeseeable hundreds of millions of years back.

Simon Conway Morris and others challenge this 'evolution is fraught with radical contingency' argument by claiming that there is good evidence for 'convergent evolution,' a kind of inevitability that evolution will hit upon certain solutions to problems faced by species of certain kinds. (Among the considerations they marshal is the fact that certain structures such as the eye have evolved multiple times, independently.) With regards to our meteor scenario in particular, Conway Morris argues that mammals were already on a path of biological ascendency and, even without the radical climate and vegetation change, would have steadily developed much as they in fact did. Since this is a debated biological issue, let me here set aside the significance of biological convergence and assume the truth of the radical biological contingency thesis.

As the very example of the meteor so dramatically illustrates, biology is not a 'complete' science in the sense that non-biological factors play an important role in explaining biological outcomes. (A more everyday example is provided, unfortunately, by auto accidents.) The biosphere is set within a wider physical realm. What is more, the structures and mechanisms of the species populating the biosphere are ultimately grounded in explanatorily and ontologically more basic physical structures and mechanisms. As a result, what looks radically contingent from a narrowly evolutionary-biological point may be inevitable from the point of view of LaPlace's physical calculator—or God. If there is a problem with seeing God as exercising providential control over the evolutionary development of humans, it will have to come from physics, not biology alone.

As it happens, physics does raise such a challenge, in the form of indeterminism in the processes described by quantum mechanics, our most basic science. Well, the proper interpretation of the spectacularly successful formalism of quantum mechanics is, famously, anyone's guess. On one empirically adequate interpretation, indeterminism is merely epistemic, not ontological. But let us suppose that there is genuine, ontological indeterminism down in the basement of our universe. Suppose further (what is not inevitable) that this indeterminism manifests itself at the level of description appropriate to genetic mutations. Then it is quite possible that mutations entered the gene pool of our ancestors in the very distant past at crucial junctures, influencing evolutionary trajectories that gave rise to us. If all this were so, it might be that not even an omniscient being could have foreseen and planned on that basis for the particular outcome of the physically indeterministic events constituting that crucial mutation or mutations. There is much to say (and which has been said, by philosophers) relevant to the last step in the argument, occurring in the previous sentence. But one thing a theist may say is simply this: if all this were so, there is nothing in science that precludes our supposing that God supernaturally insured that the 'right' (i.e., desired) mutation occurred. What was physically undetermined was made to occur by the will of God.

At this point, critics of a certain temperament will cry, "God of the Gaps theology! I knew you were going to resort to that scurrilous, anti-scientific tactic at some point!" To which I reply, "Calm down. If this is God intervening in certain 'gaps,' it is not of an objectionable sort—not objectionable unless one has already ruled out a priori that the world is caused and sustained by God. If that's your view, it's a philosophical, not scientific commitment that is driving your argument." I am no fan of God-of-the-gaps moves in natural theology, where God is invoked as the immediate cause of some as yet not-well-understood natural phenomena or process in such a way as to preclude there being a natural form of explanation. The sorry history of such moves is a large part of the reason that many educated people see a tension between science and Christian theology. ("How do we account for evolutionary transitions across major phyla? God created ex nihilo some initial set of creatures for each such phyla.") But nothing like that is being entertained here. Ex hypothesi, on some rare occasions there are outcomes of basic physical processes that are physically undetermined, and these outcomes determine what form a particular mutation will take, a mutation that is a necessary condition on a large-scale evolutionary trajectory's occurring as it does. We do not know that events meeting this description actually occur, we are supposing that they do for the sake of argument. What I (and others) suggest in response to this conjecture is that what the physical processes leave open on such occasions, God selects and determines a particular outcome. The selected outcome is consistent with what the physical laws predict, and so God's special activity is not posited as an alternative to a possible naturalistic explanation. LaPlace's calculator, observing such events, would be none the wiser, as the event would fit the patterns hitherto observed for similar events.

Space precludes giving full discussion to issues this sort of suggestion raises concerning the nature of proper scientific commitments. I will have to say, somewhat cavalierly, that we need to be ever mindful of the difference between science proper and metaphysical glosses on science. Newton was conscious of such a difference when he famously said in the "General Scholium" appended to his *Principia*, "I feign no hypotheses" concerning the basis for the properties of gravity posited by his theory. (Indeed, Newton was instrumental in cementing the difference by shaping the way that we have come to understand scientific explanation.) Following philosopher David Hume, some scientists and philosophers see in the general laws and mechanical descriptions of various sciences nothing but contingent patterns running among distinct natural occurrences. They are simply compact, observationally adequate descriptions of the regular ways that events occur. Others, going all the way back to Aristotle, see more: the best interpretation of the success of mature sciences is that their theoretical descriptions approximate entities and structures disposed towards just the outcomes observed. That is, there is a kind of natural necessity to the way that physical objects and structures behave. Either of these views, I take it, are fully consistent with science proper. Consider the view suggested by Hume. If that minimalist view is consistent with the achievements of science—things 'just happen,' and happily for us, they just happen to happen in regular patterns, so we can do science, there's nothing more to be said—then so is the following twist on it: everything that happens happens because God directly brings it about (God is its sole and complete metaphysical cause), and happily for us, God chooses to do so in regular patterns, so we can do science. That is the view known as 'occasionalism.' Now, my point in this brutally but necessarily brief review of some disputed metaphysical glosses on science is simply this: scientific commitment is, first of all, commitment to using, or respecting the use of, certain methods to study such things as natural processes, patterns (both developmental and recurring), and important limit phenomena such as the origin of the universe or of life. And it is, second, commitment to accepting provisionally the theories, laws, mechanisms, and descriptions that are well-confirmed outcomes of using the approved methods. Having made those commitments, the metaphysician, whether religious or not, is free to work within the constraints they impose to construct a vision of reality. Whether her vision is reasonable will be adjudicated on philosophical, not scientific grounds.

Another challenge that modern biology poses to traditional Christian theology concerns the theological role played by the story of Adam and Eve, especially in much of Western Christian thought. According to much of Western Christian theology, theological truths embedded in this narrative include the proposition that human beings were made for fellowship with God and originally experienced such fellowship until certain of them became estranged from God through a form of prideful rebellion. Rather than immediately repairing that breach, God chose to allow humans to continue in their estrangement but made plans to redeem them in a way that requires the free cooperation of human individuals. At least as traditionally interpreted, Adam and Eve are depicted in Genesis as the first pair of human individuals. But modern genetic analysis indicates that the human population was never smaller than between 2,000-10,000 individuals. So, whence 'the Fall'?

Here, science has indeed forced reasonable Christians to re-think an important Christian teaching. While it may come as news to many, even traditional Christian theology at its best has been a progressive tradition of inquiry, anchored by creedal fixed points. There have been a number of creative proposals for how one might envision a 'fall' event or process fully consistent with what biology reveals about human history. (To be fully consistent, they of course must not include the common but unnecessary assumption that animal predation and suffering, and human proclivity for immoral behavior of various kinds, all happened post-'fall.') These proposals are necessarily speculative, as we lack concrete data (whether from natural history or revelation) that would enable such proposals to be put to the test. They are and will remain 'just-so' stories, perhaps somewhat adjudicable for goodness of fit with biological evidence and theological doctrine, but hardly decisively. Rather than spell out these alternatives here, I refer the reader to Christian biologist Dennis Alexander's fine book, *Creation or Evolution: Do We Have to Choose?*, which canvasses several of them.[1]

My mention of animal predation and suffering brings us to a final, very widely voiced challenge to theology that is often seen as made especially acute by our understanding of evolutionary history. It is the fact of widespread, intense, unjust, and indeed seemingly random experience of great pain and suffering, followed eventually by death, itself often gruesome. Though the challenge here is quite significant, I will be very brief. I do so in part because modern philosophical discussion of the 'problem of evil/suffering' is quite complex, and in part because I don't see that modern science really adds much to the force of the problem. Even if the pre-human natural history of animal suffering were not as extensive as it has been shown to be, we have more than enough animal and human suffering within the human era to pose the problem. Yes, evolutionary biology takes away the option of chalking it all up to the disastrous consequences of human misuse of free will, but that was already woefully inadequate for an adequate explanation of the suffering that humans have been able to observe.

So what alternative form of theodicy do I propose? None. I am unaware of any fully adequate account of why an omniscient, omnipotent, omnibenevolent Creator would allow all of the kinds of horrendous suffering that occurs in and around us. But I also don't take the inability of human beings to come up with such an account to be significant evidence against the claim that God exists and has morally good reasons for permitting such suffering. It would be such evidence only if it were plausible that if there were such a reason, we would be able, in time, to discern it. But we do not have good reason to suppose this. Our perspec-

---

[1] Monarch Books, 2008

tive is too limited, our grasp of the complex ways that great good and horrible evils can interact is too meager, and, perhaps most importantly, we have no reason to believe that we are in a position to see, even dimly, all of the kinds of great goods that eschatologically transformed human nature is capable of experiencing. Christian theology is committed to a human afterlife (on which more anon). Apart from such a commitment, the facts of unjust suffering would indeed constitute the materials of a compelling argument against theism. But given such a commitment, all bets are off when it comes to whether such facts, including those revealed by natural history, might play an integral role in a future life of surpassing value.

I am aware that these brief remarks will seem trite and callous to many offended by the idea of theism in the face of human and animal suffering. They certainly are not words that would provide (or are intended to provide) comfort to a terribly suffering person. Even so, they seem to me to be true and to pose a significant obstacle to those who would give, not merely an impassioned speech intended to cow the theist into silence in the manner of Ivan vis-a-vis his brother Alyosha in Dostoevsky's *Brothers Karamazov*, but an argument with clearly articulated premises from suffering to the (probable) non-existence of God. The reader is encouraged to consult the extensive recent philosophical literature on this point under the heading of 'skeptical theism.'[2]

Psychological and Brain Sciences and the 'Soul'

In much of popular religious (indeed human) thought, human beings are or have as a part an immaterial soul that is the bearer of psychological attributes. It is this, rather than anything bodily, that constitutes our identity as persons and that enables us to survive death.

While this mind-body dualism can seem very natural from a first-personal point of view as an experiencing subject, and it is ably defended in the writings of contemporary philosophers Richard Swinburne and Dean Zimmerman, among others, I take it to be implausible given the explosion of information coming from the third-person perspective of the natural sciences, specifically evolutionary and developmental biology and cognitive neuroscience. This information, while still incomplete and only imperfectly understood, sheds light on the deep natural history of humans and present-day animals; the processes by which individual organisms of any species develop from inception to maturity; some of the function-specific neural structures and processes that sustain and help regulate the unfolding first-person perspective of

---

[2] See Trent Dougherty and Justin P McBrayer, eds., *Skeptical Theism: New Essays.* Oxford University Press, 2014

conscious agents; and finally, observed correlations between increasing complexity of neural structures and increased psychological complexity (in organismic development and across sentient species). This third-personal scientific information does not comport well with the two-substance or dualist metaphysical account of human persons. The fundamental problem is that our sciences point to highly continuous processes of increasing complexity, but the two-substance account requires the supposition of abrupt discontinuity. The "coming to be" at a particular point in time of a *new substance* with a suite of novel psychological capacities would seem to be a highly discontinuous development, both in large-scale bio-geological time and within the development of individual organisms. Furthermore, since souls as purely immaterial things would lack parts, we cannot make sense of the accumulation or diminishment of capacities by proposing increased or decreased structural complexity within the bearer of such capacities. And it just seems implausible to suppose that all the necessary basic capacities for, say, calculus problem-solving are there in the human soul from the beginning, awaiting only physical maturation in the body in order to become activated, rather than being directly dependent on that maturation for their very existence. It seems rather that psychological capacities arise and develop in tandem with the development of the brain and nervous system.

Of course, it is possible for the soul-body dualist to retrench: we might offload to the brain 'side' of the divide some of the psychological functioning that, prior to the advent of neuroscience, we might have mistakenly thought belonged to the soul. But that tack risks (as neuroscience progresses) reducing the soul to a simple, immaterial object that is radically incomplete, a mere "bearer of consciousness" that enables personal identity over time and through death.

Now, among the very many who agree with what I have just said, it is common to embrace the opposite extreme, on which conscious states are either epiphenomenal—having no influence on other psychological states or bodily behavior—or (somehow) consist in complex states of the brain. Yet this seems to me to be even less plausible than mind-body dualism. Conscious states of experience, thought, emotion, and purposive agency are our most immediately accessible empirical phenomena, and consequently they lie at the root of all our understanding of the world around us. We are not simply given *the world* to our understanding, we are given most immediately our *experiences* of it. To deny this givenness is to cut off the branch on which scientific understanding sits. And, while not self-defeating, the claim that all such experiential and belief states and purposive intendings just are complex neural states is also deeply implausible. We have direct, first-personal acquaintance with properties of these states that are manifestly different in kind from the complex, hierarchically-structured, physico-chemical properties of the brain states that are the most plausible candidates for such an identification. Just consider the feeling of a sharp pain or of coming to understand a complex scientific idea; the look of a red rose in bright sunlight; the confident, considered belief that Beijing is the capital city of China; the thought that it is doubtful that there is life on Mars; and your conscious decision to pick up some milk on the way home. Each of these states have distinctive, immediately apprehended intrinsic features that in no way resemble the sorts of features had by complex neural states on our best theories. True, recent findings have shown the fallibility and manipulability of our conscious self-awareness. Here, it will suffice to observe that the inference from fallibility to worthlessness is a poor one and one that plays into the hands of the radical skeptic concerning any human knowledge of the physical world itself. (Just run the inference on our fallible senses and inferential capacities.)

What does a middle way between dualism and reductionism look like? Among the many terms that have gained currency, "emergentism" is perhaps the most popular. But we should be careful to note that this term has meant different things to different thinkers. Here I shall use the term to indicate a view of the natural world on which human persons and other sentient animals (and *possibly* a wider array of impersonal complex systems) have irreducible and efficacious system-level features. These features are originated and sustained by organizational properties of the systems (in animals, by properly functioning brain and nervous systems) while also having in turn a causal influence on components of the system. That is, emergent systems involve an interplay of 'bottom-up' and 'top-down' causal factors. There can be—in principle—no adequate description of such a system simply in terms of the outworking of fundamental physical forces in and around it. Emergent properties are primitive features of complex entities that make a fundamental (non-redundant) difference to the way the world unfolds. They confer a substantial unity on the systems, such that one is *required* to treat them as wholes in any minimally adequate characterization of the character and dynamics of the world.

While emergent systems are not fundamental building blocks of the world, they are, so long as they persist, causally basic entities. Why is this important to insist upon in connection to religious (and specifically Christian) belief? One reason is that it is consistent, in a way that the austerely reductionist picture is not, with Christian teaching that the category of person is of fundamental significance and that human beings are capable of moral freedom. The personal does not reduce to the impersonal, and mature, intact human beings are capable of making choices in a way that confers moral responsibility.

A second reason for insisting on the unity or basicality of persons, despite their materially composed nature, is that it better comports with the teaching that we will all survive death. This will not be obvious. For emergentism, too, offers an embodied view of the soul, and we all know what death entails for our bodies. Note that in all the Abrahamic religions, human persons are not naturally immortal. (All of created reality is sustained in existence by God.) Survival of death would be a supernatural gift. Alas, we don't (yet) get to see the miracle in action of God's transporting us into another form of life and we have not been vouchsafed an account of how it goes. Thus, all we can do is speculate. Philosophers have done plenty of that, but there's no space here to survey some of their ingenious ideas. But, to get you started, note that no particular bits of matter are essential to any living thing—biological life is continual change. On the emergentist account of embodied persons, I suggest, what survival would require is sufficient psychological continuity embodied in a minimally materially continuous but changing process. And with that, I offer a teaser: if God could endow the particles of my body (or some crucial subset of them) with the ability to fission into separated spaces, and arrange for this to occur just at the moment of my demise, then maybe...

3. Some theorists maintain that science and religion occupy non-overlapping magisteria—i.e., that science and religion each have a legitimate magisterium, or domain of teaching authority, and these two domains do not overlap. Do you agree?

My discussions above of the relation of biological and psychological and brain sciences to elements of Christian theology suffice to indicate that both in fact speak to overlapping phenomena: human nature. And since I accept that both have things to teach us, I cannot agree with the 'non-overlapping' part of the claim, taken strictly. And those who reject Christianity and any other religion purporting to speak to such matters will reject the 'religious magisterium' bit. Really the only ones who can accept the claim as its stands are advocates of a neutered theology.

However, the suggestion (by Stephen J. Gould) is often cited because it is in the neighborhood of something true: the questions to which science and a religion such as (un-neutered) Christianity propose answers are to a very significant *degree* non-overlapping. Christian theology to a great extent is concerned with our 'knowing' and being rightly related to God and to the ways that our relationships with one another should flow from such knowledge and relationship. God and God's purposes are simply not an object of empirical science.

4. What do you consider to be your own most important contribution(s) to theorizing about science and religion?

My main contribution is my 2008 book *Theism and Ultimate Explanation*.[3] In it, I argue for the explanatory power of theism with respect to the question of why this particular physical reality we inhabit exists. I argue that the explanation it provides complements, rather than competes with, current and future attempts to better understand the basic structure and dynamics of the universe and its physical origins. Theism, as I see it, contributes to a more intellectually satisfying, because more complete, account of reality.

I am currently writing a book that more squarely addresses specific issues concerning the integration of our best current scientific understandings (from both particle and spacetime physics, evolutionary biology and its more speculative recent offshoots, and the psychological and brain sciences) into a broader Christian understanding of reality. As my brief remarks here indicate, I believe that a non-ad hoc, intellectually satisfying integration is achievable.

5. What are the most important open questions, problems, or challenges confronting the relationship between science and religion, and what are the prospects for progress?

One open question is whether extraterrestrial intelligent, morally-governed life can be harmonized with the Christian doctrines of the incarnation and atonement. (For those who take multiverse hypotheses in physics seriously, such non-human life is virtually certain and incredibly extensive.) I have explored this matter in a forthcoming paper co-authored with Philip Woodward.[4]

Another matter that has been discussed but perhaps not resolved is whether certain kinds of naturalistic explanations of the human disposition to religious and moral belief, experience, and practice call into question beliefs in the objective reality of God or morality. It is one thing to explain, another thing to explain in a way that appears to debunk—to reveal the practice, somehow, as not rooted in response to the putative reality that is the object of the beliefs or experiences. What kind of candidate explanations are debunking? This is much harder to say than many in the commentariat seem to assume, but it seems that some are. (Freud's account of theistic belief as wish fulfillment has little to recommend it, empirically. But it certainly seems to be debunking.)

---

[3] Blackwell Publishing.

[4] "Trans-Universe Identity: Incarnation and the Multiverse," in Klaas Kraay, ed., *Theism and the Multiverse*. Routledge Studies in the Philosophy of Religion. Routledge Press, 2014.

I advise scientists to leave this question to philosophers who specialize in epistemology. Though it's rather more popular, amateur philosophy is no less cringe-inducing to the professionals than is amateur science!

19

Massimo Pigliucci

Massimo Pigliucci is Chair of the Philosophy Department at Lehman College and Professor of Philosophy at the Graduate Center of the City University of New York. He has a Doctorate in Genetics from the University of Ferrara (Italy), a PhD in Evolutionary Biology from the University of Connecticut, and a PhD in Philosophy from the University of Tennessee. His research interests include the philosophy of biology, in particular the structure and foundations of evolutionary theory, the relationship between science and philosophy, the relationship between science and religion, and the nature of pseudoscience. Pigliucci has published over a hundred technical papers in science and philosophy. He is also the author or editor of ten books, both technical and public outreach, including *Denying Evolution: Creationism, Scientism and the Nature of Science* (2002), *Making Sense of Evolution: Toward a Coherent Picture of Evolutionary Theory* (with Jonathan Kaplan, 2006), *Nonsense on Stilts: How to Tell Science from Bunk* (2010), *Answers for Aristotle: How Science and Philosophy Can Lead Us to a More Meaningful Life* (2012), and most recently *Philosophy of Pseudoscience: Reconsidering the Demarcation Problem* (ed. with Maarten Boudry, 2013). Pigliucci has been awarded the prestigious Dobzhansky Prize from the Society for the Study of Evolution and has been elected Fellow of the American Association for the Advancement of Science "for fundamental studies of genotype by environmental interactions and for public defense of evolutionary biology from pseudoscientific attack." He pens the Rationally Speaking blog and co-hosts the podcast by the same name.

---

1. What initially drew you to theorizing about science and religion?

It's a long story. It began in 1995, when I moved to the University of Tennessee in Knoxville, as a freshly appointed Assistant Professor of Evolutionary Biology. Up to that point I had never written about science and religion, though I had been an atheist since my high school days in Rome, Italy. A few months after I arrived in Knoxville I opened the paper and found out to my astonishment that the Tennessee state legislature was seriously considering a bill that would have mandated equal time for the teaching of creationism in the local public schools. I was flabbergasted, and suddenly the meaning of the term "Bible Belt" hit me, along with the recollection that Knoxville is only a short drive away from Dayton, TN, where the famous Scopes "monkey" trial took place back in 1925.[1]

Together with some of my colleagues and graduate students I got into

---

[1] See *Summer For The Gods: The Scopes Trial and America's Continuing Debate Over Science and Religion*, by E.J. Larson, Basic Books, 1997.

action, writing to legislators and to the papers. Once the emergency was over (the bill never made it out of committee), I thought it would be better to take a pro-active stance about the whole evolution-creation "controversy," rather than just going back to the lab as if nothing happened while the next crisis was brewing somewhere nearby.

So we started one of the first Darwin Day celebrations,[2] back in 1997, which soon got me to debate creationists, give public lectures about the nature of science, and write articles and books about it.[3] As of this writing my professional interests (the nature of science and pseudoscience) and my outreach activities have pretty much converged, and I write about these matters equally for technical and lay audiences.[4] I must admit that that initial shock in Knoxville has changed my life for the better, adding meaning to it. I have made good friends along the way, and I feel like I'm not just an academic locked up in the Ivory Tower, but one who doesn't mind getting his hands dirty in public debates that matter.

2. Do you think science and religion are compatible when it comes to understanding cosmology (the origin of the universe), biology (the origin of life and of the human species), ethics, and/or the human mind (minds, brains, souls, and free will)?

The short answer is: no, on all counts. But let me elaborate. The question essentially asks whether religion has any credible epistemic authority in three areas: empirical understanding of the world, morality, and metaphysics. It seems to me beyond reasonable doubt that the first area is best served by science, while the other two are best understood as part of the practice of philosophy (with one caveat to which I'll get in a moment). This, of course, is only a first approximation, since science cannot be done without a number of philosophical assumptions, and philosophy in turn cannot be done without science setting constraints and providing necessary factual and theoretical knowledge.

To unpack my position a bit, let me start with our empirical understanding of the world. It seems to me indubitable that religion, unlike science, does not have and could never have any means to reliably discover anything at all about how the world works. Religion consists of a mixture of superstition, mythology, and folk understanding. Indeed, religion's record in this respect is simply abysmal. Creation stories from around the world have one thing in common: they are all

---

[2]   http://www.bio.utk.edu/darwin/

[3]   See in particular: Denying Evolution: Creationism, Scientism, and the Nature of Science, by Massimo Pigliucci, Sinauer, 2002.

[4]   See: Philosophy of Pseudoscience: Reconsidering the Demarcation Problem, ed. by Massimo Pigliucci and Maarten Boudry, University of Chicago Press, 2013. See also a number of entries at www.rationallyspeaking.org

false. Granted, science doesn't give us "the" Truth, only the best understanding that human beings can aspire to. But that's resulted in an impressive track record (despite the occasional blunder and dead end[5]) to which religion simply doesn't have any counter, at all. It's not just that the Judeo-Christian-Muslim "account" of creation doesn't have anything to do with what actually happened, it's also that alleged parallels between the intuitions of certain religions (say, Buddhism, which at any rate hardly qualifies as a religion to begin with) and modern science are at best vague and at any rate would have remained entirely unsubstantiated had science not provided us with the means of empirically verifying our intuitions. And let's not even speak of the creation myths of early polytheistic religions, such as the Greek-Roman ones.

When it comes to the second area, morality, the discussion becomes a bit more complicated, because folk wisdom does go a good way toward helping us with ethical questions, and if religions are, to a point, a distillation of folk wisdom then they can certainly be useful in this respect. The problem, of course, is the alleged source of moral authority claimed by religions: one or more supernatural entities who simply dictate what is right or wrong. Here I'm with Plato. In his Euthyphro[6] he had Socrates brilliantly argue that even if the gods exist they cannot possibly be a good source for morality. This is because of the famous dilemma that takes its name from the main character in the dialogue: to get morality from gods means either that one is making an appeal to (divine) authority, which reduces morality to a matter of might makes right; or that even gods ultimately must appeal to some external sense of right and wrong, a sense that should be accessible to mere mortals as well, thus rendering the gods superfluous. I am, of course, aware that countless theologians have attempted to refute Plato on this, but in my opinion they have all failed abysmally,[7] in some cases apparently without even understanding his argument.

Moreover, we have a very solid alternative to religion when it comes to ethics: moral philosophy. Philosophers have been preoccupied with ethical questions for literally millennia, and contra common wisdom they have made a lot of progress concerning them. For instance, we now have at the least three well articulated broad frameworks for thinking about morality: secular deontology (Kant),[8] utilitarianism (Mill),[9]

---

[5] Scientific Blunders, by R. Youngson, Carroll & Graf Publishers, 1998.

[6] See http://plato.stanford.edu/entries/plato-ethics-shorter/

[7] See Chapter 18 of Answers for Aristotle: How Science and Philosophy Can Lead Us to A More Meaningful Life, by Massimo Pigliucci, Basic Books, 2013.

[8] http://plato.stanford.edu/entries/ethics-deontological/

[9] http://plato.stanford.edu/entries/utilitarianism-history/

and virtue ethics (Aristotle).[10] All of them have been constantly refined and applied to real situations affecting our lives. Just consider the role of biomedical ethicists in many modern hospitals, or the fact that cadets at West Point military academy are taught about virtue ethics and trolley dilemmas.[11] Or think about the huge influence of philosophers like John Rawls and Peter Singer in modern times. The list could go on for quite a while.

Finally, let me get to metaphysics, insofar questions about mind, free will and the like are concerned. Metaphysics is, of course, a branch of philosophy, albeit a more controversial one than, say, ethics, since some positions held by notable metaphysicians are difficult to separate from the sort of incomprehensible mysticism that is typical of religious metaphysics (e.g., I'm not sure that Heidegger's concept of Being makes a heck of a lot more sense than the Christian idea of transubstantiation). Modern analyses of free will,[12] with the debates between different schools of compatibilism and incompatibilism, are a lot more intellectually sophisticated and interesting (granted, if you are into that sort of thing) than anything proposed by theologians.

But it is also important to realize that there is a crucial debate going on these days among metaphysicians themselves, one about what one can only call meta-metaphysics, and that is very pertinent to the issue at hand. The disagreement is between defenders of what can be termed classical approaches to metaphysics[13] and proponents of what is often called naturalized, or "scientific," metaphysics.[14] The debate hinges on the role of science in metaphysics: the first camp sees science as essentially irrelevant to metaphysical inquiry, while the second group thinks that doing metaphysics without close connections to science is no longer tenable, if it ever was. I myself fall somewhere in the middle, since I think there is a continuum of metaphysical issues, some of which can benefit more and some less from input from the natural sciences. The point is that the field as a whole is very much alive and kicking, again in stark contrast to what I see coming out of theology. No surprise there, of course: if one's basic axiom in doing metaphysics is that there exists a transcendental world populated with one or more supernatural entities then one is very much off to a pointless start.

---

[10] http://plato.stanford.edu/entries/ethics-virtue/

[11] http://www.philosophyexperiments.com/fatman/

[12] http://plato.stanford.edu/entries/freewill/

[13] Metametaphysics: New Essays on the Foundations of Ontology, ed. by D. Chalmers, D. Manley and R. Wasserman, Oxford University Press, 2009.

[14] Scientific Metaphysics, ed. by D. Ross, J. Ladyman and H. Kincaid, Oxford University Press, 2013.

And of course some of the examples mentioned in the question (brain and mind in particular) fall at the intriguing borderlines between philosophy of mind and cognitive science, which together represent one of the best current examples of bridges between the sciences and the humanities. Again, though, even here nowhere do I see any contribution from theology worth pondering more than a few minutes in the interest of peaceful academic relations and common courtesy.

3. Some theorists maintain that science and religion occupy nonoverlapping magisteria—i.e., that science and religion each have a legitimate magisterium, or domain of teaching authority, and these two domains do not overlap. Do you agree?

No, I don't. The chief reason for it is that I think—as I have argued above—that the "magisterium" (to use Stephen Gould's famous phrasing[15]) of religion is actually empty. Religion has no authority when it comes to an understanding of the natural world, as Gould himself of course stressed. But it has also no authority in the realm of morality, contra Gould's somewhat naive and/or Pollyannaish view. There really is very little else to be said about it, I think.

4. What do you consider to be your own most important contribution(s) to theorizing about science and religion?

Good question. I don't know whether "important" is the appropriate word here, but I will mention two, one broad, the other more specific. My broad contribution, such as it is, can be found in the corpus of my writings aimed at the general public, particularly my *Denying Evolution: Creationism, Science and the Nature of Science*, as well as in *Answers for Aristotle: How Science and Philosophy Can Lead Us to a More Meaningful Life*.

In the first book I provide an analysis of the broad conflict between science and religion, with specific reference to the American cultural wars concerning the teaching of creationism and evolution. I interpret those clashes in terms of a long and well documented history of (partly religiously fueled) anti-intellectualism in the United States,[16] as well as in terms of substantial public misunderstandings about the nature of science itself (misunderstandings in part fostered, unfortunately, by the way science is presented in text books and in a significant portion of the popular science literature).

---

[15] See: Rocks of Ages: Science and Religion in the Fullness of Life, by S.J. Gould, Ballantine Books, 1999. And also my critique of it in: Durm, MW and Pigliucci, M. 1999. Gould's separate 'Magisteria': two views, Skeptical Inquirer 6:53-56.

[16] Anti-Intellectualism in American Life, by Richard Hofstadter, Alfred A. Knopf, 1963.

The second volume is more of a self-help book for people who don't believe in self-help books, so to speak. It begins by acknowledging that human beings seek answers to universal questions, about what makes their life meaningful, the role of friendship and love, the nature of morality and justice, and so forth. I then argue that religion, the classical source of answers for these kinds of problems, actually provides nothing of the sort, and that it is far better to turn to a combination of science (providing us with the best empirically-based knowledge of how the world works) and philosophy (which gives us the critical thinking, analytical tools to reflect on what we do and why we do it).

The more specific contribution I have to offer comes in the form of a paper I published in Science & Education,[17] entitled "When science studies religion: six philosophy lessons for science classes." In it, I argue that it is an unfortunate fact of academic life that there is a sharp divide between science and philosophy, with scientists often being openly dismissive of philosophy, and philosophers being equally contemptuous of the naïveté of scientists when it comes to the philosophical underpinnings of their own discipline. I then explore the possibility of reducing the distance between the two sides by discussing some interesting philosophical aspects of research on scientific theories of the origin of religion. In the paper I show in what sense philosophy is both a discipline in its own right as well as one that has interesting implications for the understanding and practice of science. The upshot, as far as the present discussion is concerned, is that a combination of science and philosophy both explain the phenomenon of religion and completely undercut any claim of religion to have authority in either empirical or conceptual (including moral) matters.

5. What are the most important open questions, problems, or challenges confronting the relationship between science and religion, and what are the prospects for progress?

At the risk of sounding flippant, I do not think there are any open questions, aside from the (practically hugely important) ones of determining why religion persists and how to alleviate or diminish as much as possible its influence in society.

In this sense I treat religion as a form of pseudo-philosophy, just like, say, homeopathy is a form of pseudo-science.[18] We know homeopathy doesn't work as a type of medical practice, and yet many people keep

---

[17] When Science Studies Religion: Six Philosophy Lessons for Science Classes, by Massimo Pigliucci, Science and Education 22 (1):49-67, 2013.

[18] Nonsense on Stilts: How to Tell Science from Bunk, by Massimo Pigliucci, University of Chicago Press, 2010.

using it, wasting financial resources and at least occasionally jeopardizing their own health by foregoing more efficacious remedies for their ailments. We also know, of course, that homeopathy does have some beneficial consequence, subsumed by the placebo effect; but we judge this positive aspect to be more than countered by the negative ones.

Similarly with religion. Yes, it does tap into many people's need for transcendence, a sense of community, and moral guidance. And religion has had positive effects throughout history, largely by inspiring many to do good things for humanity (it has, of course, also fostered or facilitated all sorts of horrific deeds). But we have likewise solid reasons to think that there is no fundamental truth to the claims of religion, and that we have ways of getting its benefits by way of other practices (i.e., deep meditation and secular humanism social practices and philosophy).

So the way I see it the challenge moving forward is to find more effective ways to reduce the influence of religion in favor of philosophy (broadly construed, not in the narrow sense of the modern academic discipline of that name), just like the challenge posed by pseudoscience is to find ways for people to abandon it in favor of the more solid, albeit of course always revisable, view of the world offered by science.

20

John Polkinghorne

John Polkinghorne, KBE, FRS, is an English theoretical physicist, theologian, and Anglican priest well known for his contributions to explaining the relationship between science and religion. He was Professor of Mathematical Physics at the University of Cambridge from 1968 to 1979, when he resigned his chair to study for the priesthood, becoming an ordained Anglican priest in 1982. In 1986 he returned to Cambridge as Dean of the chapel at Trinity Hall, and in 1989 he was named President of Queens' College, a position he held until his retirement in 1996. Polkinghorne is the author of five books on physics and over twenty books on the relationship between science and religion, including *The Quantum World* (1989), *The Faith of a Physicist: Reflections of a Bottom-Up Thinker* (1994), *Belief in the Age of Science* (1998), *Quantum Physics and Theology: An Unexpected Kinship* (2005), *Exploring Reality: The Intertwining of Science and Religion* (2007), and *Questions of Truth* (2009). He was knighted in 1997 by Queen Elizabeth II and in 2002 received the Templeton Prize for "progress toward research or discoveries about spiritual realities."

---

1. What initially drew you to theorizing about science and religion?

I grew up in a Christian home and I cannot recall a time when I was not in some way part of the worshipping and believing community of the Church. Of course, as I grew up my thinking became more complex, but, while I sought honestly to face occasional doubts and questionings, I never felt that I was confronted by the prospect of having to deny the basic truth of Christianity.

I was a bright little boy, particularly good at mathematics which I, therefore, chose to study when I went to Cambridge University. During my undergraduate days I became very interesting in the way that beautiful mathematical equations provided the key to understanding the deep structure of the physical universe, a foundational conviction of the great theoretical physicist, Paul Dirac, whose lectures I attended. Consequently, when I came to do a Ph D, I did so in theoretical particle physics, under the supervision of the Islamic Nobel Prize winner, Abdus Salam. A research and teaching career followed, mostly at Cambridge, lasting 25 years. It was a particularly fascinating period in the development of fundamental theoretical physics, with the establishment of the quark structure of matter. I greatly enjoyed being a small party of that great enterprise and I considered it to be part of my Christian vocation to use such talents as I had in this way. I am a firm believer of the unity of knowledge and I was always seeking to understand how my scientific and religious beliefs related coherently to each other. Naturally there were some puzzles, but I never felt I faced a critical choice between science and religion. For me it was always both/and, not either/or. I saw, and continue to see, the two as friends and not foes, complementing each other in the great human quest for truth attainable through well-motivated beliefs.

In mathematical work, on the whole you do not better as you get older—it is youthful mental flexibility which counts more than accumulated experience. After 25 years of research, I felt I had done my bit for theoretical physics and the time had come to do something different. But Christianity has always been central to my life, I gradually reached the conclusion—fortunately endorsed by my wife—that my next step was to seek ordination as an Anglican priest. During my preparation for ordination, I naturally studied theology in a much more systematic way than previous occasional reading had provided. After about five years in parochial ministry, I received an invitation to return to Cambridge, initially as the Dean of Chapel at Trinity Hall. This gave me a position which combined pastoral responsibility with the opportunity for serious academic work. By this time I had reached the conclusion that, as someone with a foot in both camps, an important part of my vocation was to work on issues of the relationship between the scientific and theological perspectives on the one world which the human experience of encounter with reality is exploring. This quest for unified understanding has been my main intellectual interest for many years and I have published the conclusions that I feel I have attained in more than 20 books.

I love writing and the work of authorship has been an important aid to my thinking. Having to put ideas on paper has often served to stimulate and crystallise my thinking about science and religion.

2. Do you think science and religion are compatible when it comes to understanding cosmology (the origin of the universe), biology (the origin of life and of the human species), ethics, and/or the human mind (minds, brains, souls, and free will)?

I believe that the search for truth, attainable through well-motivated beliefs, is fundamental to both science and religion. Religious belief can guide one in life and strengthen one at the approach of death, but it cannot really do these things unless it is actually true. Otherwise it would simply be an exercise in comforting illusion. Of course, science and religion are concerned with different dimensions of reality. They ask different questions, science essentially asking 'How?' things happen, exploring the processes of the physical world, while religion asks 'Why?' is there meaning and value and purpose in what is happening.

Adequate understanding requires that both questions be addressed, so that science and religion essentially complement each other rather than being in conflict. I like to say that I am 'two-eyed,' seeking to view the world with both the eye of science and the eye of religion. I believe that this binocular vision enables me to see deeper and further than I could with either eye on its own.

For example, physical cosmology tells us the fascinating story of how the universe that we observe originated in the fiery event of the big bang, emerging from that original singularity with the initial simplicity of an almost uniform ball of energy. That simple ball of energy has now become a world of rich complexity, with ourselves the most complex consequences of the universe's 13.8 billion year history of which we are aware. This remarkable story of evolving fertility might by itself suggest that there is a deep significance in cosmic history. This possibility is reinforced by the scientific recognition that the coming-to-be of carbon-based life was only possible because the laws of nature governing the process took a very specific, 'finely-tuned,' form (the so-called Anthropic Principle). Science discovered this fact but, since it has to treat these laws simply as the given basis of its arguments, of itself it is unable to explain why this remarkable fine-tuning is the case. Yet it would surely be intellectually lazy simply to treat it as an incredibly happy accident. A religious understanding of the universe as a divine creation makes fine-tuning intelligible, for it can be understood as the endowment of the universe by its Creator of an inherent potentiality which has enabled it to have a fertile history.

In thinking about theology and cosmology, it is important to recognise that the doctrine of creation is not simply concerned with how things began but with the deeper question of why things exist at all. In other words, why is there something rather than nothing? In consequence theology must engage with the whole character of the 13.8 billion years of cosmic history. The big bang is certainly of interest, but there is no inescapable theological requirement that the universe should have been found to have a dateable beginning.

The prime characteristic of the scientific account of history, whether on the cosmic scale of stars and galaxies or on the more domestic scale of terrestrial life, is that it is evolutionary. From 1859 onwards, when Darwin had published his great work *On the Origin of Species*, there have been many Christians who have welcomed his ideas. Darwin's clergyman friend, Charles Kingsley, coined the phrase that perfectly sums up how to think theologically about an evolving world. Kingsley said that, no doubt, God could have created a ready-made world, but Darwin had shown us that God had done something cleverer than that in bringing into being a world so endowed with potentiality that creatures could explore and bring to birth its fertility as they were allowed 'to make themselves.' Such a concept is consonant with an understanding of God of love, who cannot be a kind of Cosmic Tyrant in total and unrelenting control of everything but whose gift to the objects of love must be an appropriate degree of freedom to be themselves and to make themselves.

These examples show how religious and scientific accounts are not incompatible rivals but compatible complementary insights. Some further examples may be sketched briefly. Newtonian physics, with its deterministic equations, seemed to present the picture of a clockwork universe. In such a world, creatures could be no more than complex automata and the Creator no more than the divine clock-maker who had set the machinery going and then watched it tick away. However, twentieth-century physics, with the discoveries of the intrinsic unpredictabilities of quantum theory and chaos theory, showed that the physical world is something altogether more subtle than that. A great deal of work has been done exploring what this might imply. This has made it clear that physics has by no means established the causal closure of the world on its own terms alone. It is entirely possible to take with appropriate seriousness all that science has to say and not be driven to conclude that humans have no power of free action and the Creator no room for providential interaction with creation.

Important discoveries are being made in neuroscience. Recognition of the effects of human brain damage makes it clear, I believe, that we are intrinsically embodied beings and not intrinsically spiritual beings who happen to be trapped in a fleshly body. There is clearly mutual dependence between mind and body, but neuroscience, properly evaluated, has not established their identity. A great gap yawns between talk of chains of neuronal firings and the simplest mental experience, such as feeling a toothache. No one today knows how to bridge this gap. The problem of consciousness is indeed a hard problem and talk of the triumphant armies of science being about to cross this 'last frontier' is mere bombast. While religious thinking has quite often taken a Cartesian dualist view of human nature, this is by no means essential to it and is rejected by many theologians today. In fact, the contrary view of human nature as intrinsically embodied has a long history stretching back to biblical times. Hebrew anthropology, in a famous phrase, understood human beings as 'animated bodies rather than incarnated souls.' The traditional Christian hope of a destiny beyond death is expressed in terms of the resurrection of the body rather than as a form of purely spiritual survival. We are not apprentice angels but human beings.

3. Some theorists maintain that science and religion occupy non-overlapping magisteria—i.e., that science and religion each have a legitimate magisterium, or domain of teaching authority, and these two domains do not overlap. Do you agree?

I have acknowledged that science and religion address different issues, the one asking 'How?' things happen and the other 'Why?' they are happening—is there meaning and value and purpose in what is going on? These are certainly distinct questions, but there has to be some consonance between the answers given to them. If I were to say that the kettle is boiling because gas heats the water (a How answer) and also because I want to make a cup of tea (a Why answer), that is a perfectly coherent statement to make, but if I were to say that I have put the kettle in the refrigerator for purpose of making a cup of tea, that would be a blatant contradiction! There has to be, therefore, a similar degree of consonance between what science has to say and what religion has to say. Inevitably, therefore, the two sources of insight must interact.

That this is in fact the case is clear from the actual history of the relationship between these two quests for truthful understanding. Who could deny that theology's reflection on the nature of creation and its creator was initially challenged and ultimately enriched by the scientific discovery that the universe has had a long history of evolving development? I have already suggested that the discovery of biological evolution led theologians to see that creation is not a timeless act of producing a readymade world, but the story of how creatures have 'made themselves' by exploring and bringing to birth the God-given potentiality with which the world has been endowed. This mutual influence between science and religion is not simply one-way. There is an interesting case to be made that the Judaeo-Christian concept of creation as the free act of a Creator who is reason itself (the Logos) was an important factor in the development of modern science in the seventeenth century. Because the Creator is rational there will be a deep order in God's creation, supporting the scientists' instinctive trust that the world is intelligible, possessing a structure to which the human mind can expect to gain access. Of course, the ancient Greeks also believed this, but Plato thought that the Demiurge, in forming the world, had to follow the prescribed patterns of the Forms. Because of this underlying noetic foundation, the Greeks thought it should be possible to gain access to these patterns by pure thought alone. However, the Judaic-Christian tradition stressed the freedom of the Creator to bring about whatever order he wished, unconstrained by any external plan. It was necessary, therefore, to look at nature itself by observation and experiment, to discern what God had actually chosen to do. Thus the early scientists, most of whom were

in fact religious believers, were encouraged both to look for an order
of nature expressed in rational mathematical terms and to learn from
experimental interrogation what form that order actually took, the two
foundational methodical principles on which modern investigation of
the physical world is based.

**4. What do you consider to be your own most important
contribution(s) to theorizing about science and religion?**

My work in science and religion has been shaped by my experience as
a physical scientist. This has meant (to quote the subtitle of my Gifford
Lectures), that I am a 'bottom-up' thinker, seeking to attain understanding by an initial engagement with the particularities of experience before trying to move on to an assessment of overall general significance.
This strategy contrasts with a 'top-down' approach which believes that
it can start from broad general principles ('clear and certain ideas'),
before condescending to a consideration of particulars. The basic reason for adopting the bottom-up approach is that what were considered
self-evident general principles have often proved to be neither clear
nor certain. Reality often turns out to be much stranger than everyday
experience might, at first sight, lead us to suppose. In my own subject
of physics, quantum theory makes the point most clearly. Anyone in
1899 could have undertaken to 'prove' the impossibility of an entity
to behave, sometimes like a wave (that is spread out and oscillating)
and sometimes like a particle (that is, a little bullet). Nevertheless, as
we all know, that is the way which light has actually been found to
behave. Only the stubborn nudge of the particularity of nature could
have forced the physicists to consider this oxymoronic possibility and
eventually they came to understand how it could be so. I believe that in
an analogous way, Christianity has been driven by its engagement with
the evidence of the life, death, and resurrection of Jesus Christ to embrace the concept of a deep duality of humanity/divinity in Jesus Christ.
Thus I have sought to employ this bottom-up strategy in a defence of
the reasonableness of Christianity, portraying faith, not as unquestio-
ning submission to some unchallengeable authority, but commitment to
well-motivated belief (see my Gifford Lectures).

Closely allied to bottom-up thinking is the concept of critical realism, the idea that, in science and beyond it, we make contact with an
actual reality, but one whose nature is more subtle than an Enlightenment notion of unproblematic objectivity would have led one to antici-
pate a priori. The quest for understanding must be open to our thinking
being shaped by the actual character of the reality we encounter and not
by some prior expectation (think again of the wave/particle duality of
light). Many other scientist-theologians, such as Ian Barbour and Ar-

thur Peacocke, share this stance, finding that their scientific experience encourages them to place a trusting reliance on the possibility of finding deep truths through a careful pursuit of well-motivated beliefs in many forms of human enquiry, including theology.

Within this framework of truth-seeking activity, I have been particularly concerned with a number of specific issues. One is a revived and revised form of natural theology, not seeking to rival science in its own proper domain, but the complementary activity of setting scientific insight within a broader and deeper context of intelligibility. For example, the rational transparency and evolving fertility of the universe are perceived, not simply as marvellously happy but unexplained accidents, but are interpreted as signs of a divine Mind and Purpose behind the world that science explores. This new phase of natural theology is more modest in its claims than earlier discourse. It forsakes the language of 'proofs' of God, speaking instead in terms of intellectually satisfying and persuasive insights. Atheists are not stupid, but they explain less than theists can.

Another topic which has been of great importance for me has been eschatology, the quest to understand whether there can indeed be a destiny beyond death, not only for ourselves but for the whole universe, which the cosmologists reliably predict will eventually also die, becoming, in the course of the unfolding of present process, a world of increasing old and diluteness, devoid of all forms of carbon-based life. Science by itself can do more than assert this bleak prospect, but theological belief, appealing to the faithfulness of the Creator, can affirm the hope of a divinely brought about destiny beyond death, both for individuals and for the universe itself. I have sought to explore and defend the coherence of this hope that the universe is truly a cosmos and not ultimately a chaos. Delicate issues are involved, of both continuity (for example, it must really be the same persons who live again and not new persons given the old names for old time's sake) and discontinuity (there is no point in making persons live again if they are soon to die again). In this investigation I have made use of the increasing scientific recognition of the importance of the role of 'information' in its thinking, suggesting that the human soul, considered as the carrier of continuity between the present 'old creation' and the prospective 'new creation,' is not some purely spiritual entity temporarily trapped in a physical body, but 'the almost infinitely complex information-bearing pattern' carried at any one time by our bodies. I believe that this pattern will not be lost at death but it will be preserved in the divine memory in order that it may be re-embodied in God's great act of resurrection into the life to come. This concept of the soul is something like a recovery in modern form of the ancient Hebrew concept of humans as psychosomatic unities, 'animated bodies rather than incarnated souls.'

I also participated to the animated discussions in science and religion circles in the 1990s of how to reconcile theological belief in God's action in the world with scientific accounts of causal structure. A common feature in this activity was a recognition that the presence of the intrinsic unpredictabilities in natural processes discovered by physics in the twentieth-century (quantum theory, chaos theory) have shown that the physical world is something more subtle than mere mechanism and this means that science has not demonstrated the causal closure of the world on its own terms alone. In this work I laid particular stress on the relevance of chaos theory to the discussion.

I would say that the overall trajectory of my thinking has been to seek to hold to traditional Christian insights, while taking seriously the insights of modern science. In my view, this approach requires a degree of openness to the possibility of careful reconceptualisation, but not a rejection of the essentials of Nicene Christian belief.

5. What are the most important open questions, problems, or challenges confronting the relationship between science and religion, and what are the prospects for progress?

The least thoroughly explored part of the frontier between science and religion has been that involving the human sciences, despite their obvious importance and relevance to the issues involved. Some work has, of course, been done particularly by psychologists, but it seems that a great deal still remains to be done, including a careful evaluation of the significance of the developments currently taking place in neuroscience. The latter seems to me, despite its successes, to be still at an early stage, as it concentrates on the neural pathways by which information originating in the environment is processed by the brain. The problem of consciousness remains essentially unresolved, despite some grandiose claims that this is the 'last frontier' that the triumphant army of science is just about to cross. In actual fact, there is a vast gap yawning between talk of neural firings and the simplest mental experience of feeling hungry or seeing red. In fact, it is not clear that there will ever be a fully neuroscientific account of consciousness, since conscious experience is essentially private in character, in contrast to publically accessible experience, open to experimental manipulation, which is the concern of the rest of science. However that may turn out to be, I hope that there will be much further rigorous exploration of the human science/religion frontier and that important insights will be gained in this way.

21

James Randi

James "The Amazing" Randi is an internationally known Canadian-American stage magician and escape artist, but is best known today as the world's most tireless investigator and demystifier of paranormal and pseudoscientific claims. He has received numerous awards and recognitions, including a Fellowship from the John D. and Catherine T. MacArthur Foundation in 1986, a Lifetime Achievement Award from the American Humanist Associations in 2012, and a Lifetime Fellowship from his peers at the Academy of Magical Arts. In 1993, the PBS-TV "NOVA" program broadcasted a one-hour special dealing with Randi's life work, particularly with his investigations of Uri Geller and various occult and healing claims made by scientists in Russia. He is the author of numerous books, including *The Truth About Uri Geller* (1982), *The Faith Healers* (1989), *Flim-Flam!* (1982), and *An Encyclopedia of Claims, Frauds, and Hoaxes of the Occult and Supernatural* (1995). In 1996, The James Randi Education Foundation was established to further James Randi's work—and his long-standing challenge to psychics to demonstrate any psychic, supernatural, paranormal, or occult powers or events under agreed-upon scientific testing criteria now stands as a $1,000,000 prize and remains unclaimed.

---

1. What initially drew you to theorizing about science and religion?

As a professional conjuror, I am of course very much aware of—and concerned with—reality, since my work consists of changing my audience's perception of what's "real"—but always with the goal of entertainment. I believe that my interest in the science-vs.-religion confrontation arose naturally due to my concerns that so many persons with whom I came in contact seemed to accept that these two views of reality were compatible, which in my opinion—I don't have a "humble" one— they most certainly *are not*.

Science is firmly based on evidence. Religion is loosely based on both wishing and fear. I have always felt that the source of each of these views—'way back in history'—was the naturally-arising search for cause-and-effect: what and why. Why does rain fall from the sky? 'Way back,' I speculate that some of our forebears invented a Rain God to explain that phenomenon, as they did an Earthquake God, a Lightning God, and so on. That was a simple and direct way, and all that was needed was to also design/discover how to propitiate a deity so as to have some way of earning a favor, so prayers were invented and offered...

Another way—which became what we now know as science—was to observe that rain occurred more often at certain times of the year, and that preparations should therefore be made to benefit from that fact.

That sort of thinking constituted the beginnings of science, which was of course improved from more observations, record-keeping, and planning.

Part of the inquiry surely involved noting whether the prayers worked. And noting that they didn't...

2. Do you think science and religion are compatible when it comes to understanding cosmology (the origin of the universe), biology (the origin of life and of the human species), ethics, and/or the human mind (minds, brains, souls, and free will)?

No, science and religion are not compatible on *anything*. For understanding the origins of the universe, of life, of humans, of ethics, or of free will, I look upon some of these as perhaps beyond our reasoning and probably of little impact on our understanding of how to conduct ourselves or of adding to the improvement of our behavior. The idea of free will itself is not too complex a subject; yes, we have free will, but only at the moment that we decide whether or not to push the button or to take the left fork in the road, which depends upon what we choose to do even though that choice will be made based on our total experience, education, training, and/or fears... So the answer to that question is actually yes, and no...

The "origins of the universe" inquiry is far too simplified a question, one that we cannot answer because it would call for asking whether there could be a force, entity, or agency that could cause the universe to "be"... That's beyond reasoning. However, origins for life and for our species are already in place through Darwin's excellent view, though always subject to fierce discussions and differences, of course. But if you examine an origin for the very simplest bit of mold—the fungus sort—or any other bit of life, and you will have also examined the origins of *homo sapiens*...

As for the origins of ethics, I would say that these are common-sense rules predicated upon survival needs, being designed and put in place to impress behavior that benefit the species, though I suspect that there are certainly hard-wired—evolved and inherited—rules already in place.

I see zero evidence for souls, a strictly religious artifact and a serious surrender to wishful thinking that is seen in the present widespread acceptance of talking-to-the-dead artists and attendance at silly sessions of table-bumping "séances"...

3. Some theorists maintain that science and religion occupy non-overlapping magisteria—i.e., that science and religion each have a legitimate magisterium, or domain of teaching authority, and these two domains do not overlap. Do you agree?

Most emphatically, no. I see no specific "educational" magisterium for religion, at all. Religion derives its power from promises of relief from fear of death, and certainly provides that to those who choose to accept those promises. That, to me, is not education, but deception. It can be, and it often is...

4. What do you consider to be your own most important contribution(s) to theorizing about science and religion?

I cannot personally claim to have made any specific contribution to the science/religion inquiry, since my experience of the literature is lacking, and I have no recognized authority. However, I have devoured—with delight—the works of Christopher Hitchens, Daniel Dennett, and Richard Dawkins, to name only a few. But to return to my chosen profession of conjuring, that discipline is particularly well suited to analyzing such matters as claims of ESP, speaking-with-the-dead, dowsing, UFOs, and similar matters, simply because we "magicians" use misdirection of attention, mis-statements of circumstances, and slanted reasoning, to accomplish apparent miracles. An example: the great Harry Houdini performed a very effective vanish of a living elephant at Madison Square Garden in New York City in the late 1920s by observing that the largest and only non-compressible part of such an animal is the skull. He designed a huge box-car-like container for the elephant with a massive frilly drape that could be pulled back and forth across the large open side. A handler entered from the back to join the elephant after the curtain had concealed it from view. He directed it to turn head-for-tail, then lie down, whereupon it became very flat—except for its head, which was now at the opposite end of the container. A judicious piling of the large amount of cloth and drapes at the "head" end covered that, and the now-flattened body sagged down slightly into the floor. Voila! The pachyderm was effectively gone when the curtain was drawn back...

You can see, from the description just given, that Houdini's illusion was just that, an optical effect assisted by the decorations and a little-known factor of elephant anatomy. In a similar way, you can be informed of the *modus operandi* of "mentalism" tricks whereby "psychics" can call out the names and addresses of persons sitting in their audiences, much to their genuine surprise, and even tell them the serial numbers of currency bills they carry in their wallets and purses... No, solutions to these apparent miracles will not be provided here...

5. What are the most important open questions, problems, or challenges confronting the relationship between science and religion, and what are the prospects for progress?

I honestly can see nothing *but* questions, problems, and challenges! A set of established facts—though those facts are subject to refinement, correction, or even rejection—trumps any set of rules, conjectures, wishes or hopes arrived at through doubt, fear or ignorance of The Real World, especially because these originated centuries ago and cannot undergo any improvement.

I'll quote here Isaac Asimov [1920-1992]:

> Imagine the people who believe such things and who are not ashamed to ignore, totally, all the patient findings of thinking minds through all the centuries since the Bible was written. And it is these ignorant people, the most uneducated, the most unimaginative, the most unthinking among us, who would make themselves guides and leaders of us all... I personally resent it bitterly.

I digress somewhat in order to point out what I believe to be the major flaw in our thinking process—religion—that has doomed such a large sector of the world's population to believe in superstitious nonsense. There are so many examples that demonstrate what I believe to be obvious reasons to doubt and deny beliefs in deities and the other chimeras that are so often summoned up in their support. This is a very important aspect indeed, one to which I could devote many pages. I am an atheist, and I have been an atheist as long as I can remember. I found myself surrounded by people who offered me no cogent arguments to support their beliefs in miracles, Bible stories, or various prohibitions and superstitions stated in religious writings of several kinds. For that reason—and I believe it's an excellent reason—I became what I call an "atheist of the second kind." Many atheists claim that there is no deity. I do not belong in that category because I cannot prove that there is no deity. I belong in the second category: I do not find sufficiently good evidence to convince me that a deity exists, therefore I am an atheist of the second kind.

Recent surveys have shown that atheists actually tend to know more about religion than religious persons. Alan Cooperman is associate director for research at the Pew Research Forum, an agency based in Washington, D.C. that provides information on issues, attitudes and trends shaping the United States and the world. He says:

> American atheists and agnostics tend to be people who grew up in a religious tradition and conscious-ly gave it up, often after a great deal of reflection and study... These are people who thought a lot about religion. They're not indifferent. They care about it.

Of course, I recognize that there can be a certain amount of fear and even terror for some individuals at the very thought that they might be even doubting or questioning their religious beliefs. The fact is that I have never, ever, experienced the slightest tremor in this regard. I'm proud of being an American citizen, because that's a conscious, pur-poseful, considered, deliberate decision that I made; I was born as a Canadian. I'm not as "proud" of being an atheist, because that's just the way I am; I'm also right-handed, and I take no pride in that, either, and for the same reason.

Religion is a philosophy, a stance, a set of rules, an opinion. It offers no proof, no evidence, no facts. Arguments, yes, and claims, yes, but nothing that can actually be put to the test. I'm an atheist, not an agno-stic—one who believes that these are matters beyond discussion. I find no evidence for miracles, Heaven, or Hell. The Bible, the Talmud, and the Koran are not supported by history, and all are inconsistent in many places. Does this prove that religion is "untrue"? No, of course not. My stance is that it is simply unproven. In my opinion, skeptics should be skeptical of all claims, and religion is of course included, though some prefer to omit religion from consideration. *Chacun á son goût.*

Because religions don't offer statements that can be proven one way or the other, I have not challenged most of their claims, and for that specific reason. The only exceptions have been when a matter of faith healing, rising from the dead, or any such directly stated and defined miracles or other phenomena for which there exists examinable evi-dence, has come up. Those claims are pursued, of course, and I even wrote an entire book about the subject, *The Faith Healers*, which made my personal stance on such matters, very clear.

But I'm willing to be shown, always. That's my personal position, which I feel has to be stated firmly. Other versions of my philosophy are hereby superseded...

If you're still with me, I ask you to consider just a very few of the more ridiculous tenets of the world's major religions. We have The

Rapture, where on a very uncertain date in the future, those "saved" will simply float up to Heaven and leave the Sinners behind. Another group believes that Jesus Christ visited the American Indians shortly after his execution. In yet another group, individual angels have been assigned to everyone, and in the third-largest religion on Earth, cows are—literally—worshipped. The Hindu god Devagana lives in the human anus. Christians accept virgin birth and transubstantiation. Others believe that rocks and trees have spirits living in them, and another sector of our species leaves their dead out for the vultures to devour, while others simply let their kids die because they won't allow transfusions or other medical intervention. These are firm beliefs held by the large majority of our species. With such radically different philosophies and beliefs, how can any one of these divisions of humanity ever hope to get along with—or even tolerate—any of the others?

Questions, anyone…?

22

Alex Rosenberg

Alex Rosenberg is an American philosopher whose research focuses on problems in metaphysics, mainly surrounding causality, the philosophy of social sciences, especially economics, and most of all, molecular, functional, and evolutionary biology. He is currently R. Taylor Cole Professor of Philosophy at Duke University. In 1993, he was awarded the prestigious Lakatos Award, which is awarded annually for outstanding contributions to the philosophy of science. Rosenberg is the author of over two hundred scholarly articles and fifteen books including *Microeconomic Laws: A Philosophical Analysis* (1976), *Sociobiology and the Preemption of Social Science* (1980), *The Structure of Biological Science* (1985), *Instrumental Biology, or the Disunity of Science* (1994), *Darwinism in Philosophy, Social Science and Policy* (2000), *Philosophy of Science: A Contemporary Approach* (2000), *Darwinian Reductionism or How to Stop Worrying and Love Molecular Biology* (2006), and *The Atheist's Guide to Reality* (2011).

---

1. What initially drew you to theorizing about science and religion?

I think many people start out religious by culture working on the hard-wired theory of mind that makes us look for motives even in nature. That included me. In my case, as in that of many other philosophers, once you see through religion's metaphysics, you begin to take science much more seriously as a source of answers to the questions about the nature of reality. Then you cease to concern yourself much with religion (the philosophy of religion—arguments about the existence of god—has little to do with religion).

What drew me to the limited theorizing I have absorbed about religion and the even less theorizing I have produced is the question of why religion is almost ubiquitous in human affairs across its whole history and in every culture (bar the Buddhist and zen cultures, where god plays little role). This of course is the question Dennett addresses in *Breaking the Spell* (2006). I have little to add to his answers to this question—it's a powerful meme well adapted to parasitize humans. So I have become interested in exactly what features this meme has that enables it to colonize human minds, and social groups, so effectively to create organizations and institutions that ever increasingly get people to act against their self interests, and even the interests of their communities and mankind in general.

2. Do you think science and religion are compatible when it comes to understanding cosmology (the origin of the universe), biology (the origin of life and of the human species), ethics, and/or the human mind (minds, brains, souls, and free will)?

I take it that the Abrahamic, theistic religions—Islam, Christianity, and Judaism—makes substantive claims about "the furniture of the universe" and how it is arranged. On any literal interpretation of these claims, the theisms are incompatible with cosmology, and with much more fundamental physics for that matter, especially the 2nd law of thermodynamics. It is not worth discussing whether on some latitudinarian interpretation of the claims of the sacred texts of these three religions, they might be rendered compatible with cosmology. So far as interpretation goes, any text can probably be reconciled with any other text, including its direct denial, by judicious choice among ambiguous, vague and abstract claims.

I have argued that on a widespread interpretation of theism, it is flatly incompatible with the Darwinian theory of natural selection. I take it as a central feature of theism that god created man in his own image, that is, he had us as an intended outcome of his handiwork. If so, god could not have chosen the process of natural selection that Darwin uncovered as the means to produce us. The process of natural selection is well known to be driven by objective probabilities, the same ones that drive the 2nd law of thermodynamics. This means that, in S.J. Gould's words, if you rewind the tape of life and start over at the same initial conditions as the Earth 3.5 billion years ago, the chances of human life appearing, or even sentient life, or even life, are vanishingly small. No omniscient creature would employ a method that the creature knew was unlikely to produce the outcome the creature aimed at. Accordingly, there is no way to reconcile Darwinian evolution with theism, and so no way to reconcile religion with the science which it organizes—biology.

I believe Plato demonstrated the irrelevance of religion to ethics in the *Euthyphro*. The question remains whether science is relevant to ethics in the sense of providing a warrant, justification, grounding for any particular ethical norm. While I believe that natural selection can explain many features of the core morality most people share, I hold with Hume that it cannot underwrite any ethical norm or intrinsic value. Accordingly I must conclude that religion and science are on a par in their irrelevance to normative ethics. Of course religion—theism at least—has nothing to tell us about metaethics, owing to its incompatibility with science.

How could religion provide any insight into the mind—the nature of cognition, emotion, or sensation? Only by providing data, in particular data about how the brain can be affected by significant cultural forces. Whether we have free will or not could never be adjudicated by religion. In fact the problem that free will makes for theism—reconciling omniscience with moral responsibility—is so serious that it ties theologians up in knots—cf. Calvin, while producing a large share of atheists via its casuistical solution to the argument from evil.

Broadly speaking, science and religion are not merely incompatible, but so different in their orientation—teleological v. mechanistic, that it is unlikely they could jointly shed light on any of the things mentioned in this question. Fortunately science requires no assistance from religion in dealing with these matters to the extent they submit of comprehension.

3. Some theorists maintain that science and religion occupy non-overlapping magisteria—i.e., that science and religion each have a legitimate magisterium, or domain of teaching authority, and these two domains do not overlap. Do you agree?

No, this is one of a number of things that S J. Gould claimed in a career of serious over-reaching. It would be nice if religion never made any but normative claims. Then we might be able to compartmentalize it in ways that would block any impact on the progress of science. Alas, as Giordano Bruno, Galileo and others could have testified, theism is not content to merely be a normative or ethical magisterium that doesn't overlap with science. Four hundred years on it still wants to compete. Just ask The Discover Institute and the creationists whose latest manifestation it is.

I am inclined to go further and in agreement with Steven Weinberg, argue that religion is one of the forces which throughout history has had a baleful moral impact on civilization, and for that reason should not even be granted its own normative magisterium.

4. What do you consider to be your own most important contribution(s) to theorizing about science and religion?

I suppose that the only contribution I have made is an argument that Darwinian theory and theism are incompatible owing to the role of objective chance it attributes to evolution and the difficulty theism has reconciling a world of real objective chance, in which we might not have emerged, with the commitment to god's having created us with malice of forethought.

5. What are the most important open questions, problems, or challenges confronting the relationship between science and religion, and what are the prospects for progress?

Properly understood the relationship between science and religion is adversarial. I suspect it is important in American public life that this fact not be widely recognized owing to the role of taxation in the support of science and the proportion of religious believers among the taxpayers. In the long run the hope for human survival and flourishing requires the advancement of science and so the diminishment of the role of religion in public and probably also private life. In light of the strong natural proclivity to religion among human beings, attaining anything like this requirement for the long-term advance of science is a major challenge.

23

Michael Ruse

Michael Ruse is a British philosopher of science who specializes in the philosophy of biology and is well know for his work on the relationship between science and religion, the creation-evolution controversy, and the demarcation problem in science. Currently, he is Lucyle T. Werkmeister Professor of Philosophy and Director of the History and Philosophy of Science Program at Florida State University. Ruse is a Fellow of the Royal Society of Canada and a Fellow of the American Association for the Advancement of Science. He is the author of over thirty books, including *The Darwinian Revolution* (1979), *Darwinism Defended: A Guide to the Evolution Controversies* (1982), *Taking Darwin Seriously: A Naturalistic Approach to Philosophy* (1986), *The Philosophy of Biology Today* (1988), *The Darwinian Paradigm: Essays on Its History, Philosophy and Religious Implications* (1989), *Evolutionary Naturalism* (1995), *Mystery of Mysteries: Is Evolution a Social Construction?* (1999), *Biology and the Foundation of Ethics* (1999), *Can a Darwinian be a Christian? The Relationship between Science and Religion* (2001), *The Evolution Wars: A Guide to the Debates* (2003), *Darwin and Design: Does Evolution Have a Purpose?* (2003), *The Evolution-Creation Struggle* (2005), *Philosophy After Darwin* (2009), *Science and Spirituality: Making Room for Faith in the Age of Science* (2010), and *The Philosophy of Human Evolution* (2012). He is the editor of *The Cambridge Encyclopedia of Darwin and Evolutionary Thought* (2013) and co-editor of *The Oxford Handbook of Atheism* (2013).

---

1. What initially drew you to theorizing about science and religion?

From the very beginning, even as a schoolboy, I was interested in both science and religion, although not necessarily in connection. I was brought up as a Quaker and at least until the age of 20 was a deeply committed Christian. I started science at school and in fact went to university to read mathematics. While there I discovered philosophy and that has been my passion and profession ever since. When it came time to write a doctoral dissertation I was perhaps naturally drawn to writing in the area of philosophy of science, although in those days ethics was so boring—Prescriptivism and that sort of thing—that I choose science by default in respects too. As much for practical reasons as for anything else, namely that this was an area that was not well explored, I directed my attentions to the biological sciences and ended up writing on the philosophy of biology. This led to my first book, *The Philosophy of Biology* (1973).

This somewhat naturally led me back to Charles Darwin, it is worth noting that I was working in the 1960s when Thomas Kuhn was very influential, and before long I was as immersed in the history of biology—particularly evolutionary biology—as in the philosophy of biology. This naturally led me into discussions of the relationship between science and religion. However still this was not what I would describe as a central interest in my work or indeed in my general thinking. Things however did start to change towards the end of the 1970s. It was about this time that the biblical literalists, now calling themselves "Creation Scientists," started to make a big splash. I was called upon because of my expertise to debate a number of these people and increasingly my own personal interests and expertise in the science-religion relationship started to increase. I think it was part personality—immodestly, I am pretty good on my feet and love an audience rather like vampires like virgins—and so I found I was in demand as a speaker and debater, but also a very deep Quaker sense of the need for service (and I do think fighting crude religion is service) that led me this way.

I am a pretty happy person. I have a beautiful and loving wife, five great kids, good health, and a job I love and can still do at the age of 73. I lost my faith around the age of twenty and have never again felt the need to take it up again or to replace it by some alternative like humanism. But non-believer though I may be, I don't think I was put on this Earth to enjoy myself. I have been given much and I take the parable of the talents very seriously. I must work at and return my gifts five-fold. The Beatitudes are also all important. I must give to others. Plato had it right of course. Giving is no sacrifice. The only happy person is the good person. I am not a prig—except when it comes to the obsession at American universities with sports—but if you take my vulgar and confident demeanor at more than surface level, you much misunderstand me.

Eventually in 1981 I found myself in the state of Arkansas, appearing as an expert witness for the American Civil Liberties Union in its successful attack on a new law in the state mandating the teaching of creation science. This led to a sustained interest in the science-religion relationship, one which persists thirty years later to this day. Along the way I have written a number of books pertinent to the issues.

In this context there is one thing to which I want to pay special tribute. After the Arkansas trial, I was invited to speak at the annual conference of an organization, as modest as its name is pretentious, *The Institute on Religion in an Age of Science.* This is an American-based group founded in the 1950s by a number of people who wanted to explore the science-religion relationship, basically I think it fair to say from a non-denominational, liberal, religious perspective—over the years there have been many Lutherans, Unitarian-Universalists, United Church of Christ members, and one lone Jesuit! They used to meet once a year for a week on Star Island, one of a group of islands (formerly fishing bases) off the coast of New Hampshire, at the end of July, and have a conference on various science and religious themes, for example the challenges posed by the Human Gene Project. Things have changed a bit now as old friends have retired and died, but for over twenty years I and my family used to drive down from Ontario (where I then worked) and attend faithfully. I can simply say that those weeks, ten miles out in the Atlantic, were spiritually and intellectually those I count among the most precious in my life. The friendship, the stimulus, the provocative thinking, and the paradisiacal surroundings might almost have been an argument for the existence of God. Happy hour was pretty good too!

2. Do you think science and religion are compatible when it comes to understanding cosmology (the origin of the universe), biology (the origin of life and of the human species), ethics, and/or the human mind (minds, brains, souls, and free will)?

I don't think that science and religion are necessarily compatible. Indeed quite obviously one cannot hold to a literal reading of Genesis and to modern cosmology and evolutionary biology. However I don't see in principle why science and religion should not be compatible. In the area of cosmology I think that as long as we understand the notion of a Creator God meaning not the equivalent of the Big Bang but rather as something that sustains the whole of existence, giving a reason why there is anything at all, there is no conflict between science and religion. In the area of biology I think now we have a sufficient understanding of the workings of organic bodies that even though we do not yet have a full and satisfying understanding of the origin of life, it is certainly on the future horizon.

The same I think is true of human beings at least considered as organisms. I do think that there are some knotty problems that not all religious people recognize when it comes to humans. In particular if you subscribe to any form of Darwinian evolutionary theory, you are committed to the inherent randomness of the process. This means some special attention must be paid to any religious claims about the special nature of humans, particularly if this is understood as meaning that humans in some way are a necessary part of God's creation.

I don't think that this is an unsurpassable problem. I do not myself much favor directed mutations—so-called theistic evolution—but I suspect that some kind of naturalistic account can be offered. I myself rather favor the idea of multiverses, where even if humanlike organisms don't necessarily appear in one universe, given an indefinite number they can be expected to occur eventually. This is no problem for the Christian God, for remember he is a being outside space and time.

I think already we have been able to show that we can give a naturalistic account of ethics. I myself favor some sort of position where ethics is not so much justified as explained as a natural phenomenon. This it seems to me fits nicely with a Thomistic natural law perspective. Finally there is the question of minds. Although I think brain science has shown us a great deal about the nature of the human mind and that the metaphor of the mind as a computer is still proving very powerful, I'm not sure that we have given yet an adequate understanding of sentiments. In fact I myself am inclined to think perhaps we never shall. But I don't see any reason thinking that this calls for a clash with religion, at least the Judeo-Christian religions.

It seems to me that it's perfectly open for somebody to talk in terms of souls or whatever so long as it is understood that these are not scientific concepts but religious ones. Again in the case of free will I myself subscribe to a compatibilist view, but I don't see that any of the standard philosophical approaches are antithetical to religious understanding. I do think however that one needs to be wary of using science to justify religious positions. I'm thinking here particularly of the abortion debate. Whether or not it is justifiable it seems to me consistent for a Christian to argue that abortion is always immoral. [I think also however that a Christian could legitimately argue that abortion is allowable on various grounds.]

What I do not think is consistent is for a Christian to use our understanding of human development or brain science as justification for claims about the soul. The soul is a nonscientific concept and as such stands outside science, both from the point of view of criticism but also from the point of view of sustaining evidence. So if a Christian argues that abortion is always wrong because it is destroying a soul, a Christian can argue for this, but not on scientific grounds. Conversely a critic can argue against this, but it must be done on theological or philosophical grounds. Saying that you cut open a brain and did not find a soul is not an adequate response.

3. Some theorists maintain that science and religion occupy non-overlapping magisteria—i.e., that science and religion each have a legitimate magisterium, or domain of teaching authority, and these two domains do not overlap. Do you agree?

In a way I'm inclined to agree that science and religion are non-overlapping Magisteria. However I want to qualify this in at least two ways. First obviously as I pointed out above science and religion can come into conflict. You cannot hold to Genesis and the Big Bang theory at the same time. Likewise you cannot hold to Darwinian evolutionary theory and the six days of creation at the same time. And as significantly if you take modern geology at all seriously, then the idea of Noah's flood—a worldwide deluge—is simply impossible.

Second I don't think that Stephen Jay Gould's notion of a Magisterium does the job properly. As I read what he wants to say, he argues that science tells us about the way things are whereas religion tells us only about moral issues. Obviously if this is so then they cannot come into conflict. "This is an electric chair" is not the same statement as "I think it is good to execute people on occasion." However it is clear that religions, especially including Christianity, Judaism, and Islam, want to do more than just make moral statements. They want to make statements of fact for instance that Jesus was the son of God, that God exists, and that there is the prospect of eternal salvation. These go beyond the moral.

So in these two ways I don't think that Magisterium notion works. However I see no reason why one should not have a more stripped-down religion which does not saddle itself with outdated scientific views and the prospects of this coexisting with modern science seem to me far more favorable. Basically my position is that modern science is governed by the metaphor of a machine—that is why we speak of "mechanism"—and following Thomas Kuhn and other linguists I would argue that an important thing about metaphors is that they leave certain questions open. "My love is a rose" tells you that she is beautiful—it doesn't tell you if she is any good at mathematics.

In my opinion, the mechanistic world picture tells us nothing about ultimate origins, in the sense of why is there something rather than nothing; it tells us nothing about the foundations of morality; as I pointed out it is dubious about whether it can explain the mind fully; and obviously it does not speak to any future eternal prospects. You may think that while these are all legitimate questions, they are questions that we cannot answer, and in fact I'm sympathetic to this position. But if the religious person wants to give his or her answers in terms of religion, I don't think that science can lay a finger on this practice.

If the religious person wants to say a creator God is responsible for the world, that morality in some sense is the will of God, that minds are a sign that we are made in the image of God, and that He holds out the prospects of eternal salvation, then these are not scientific claims and whether or not you accept them science as such cannot disprove them. However, as I pointed out above, where care must be taken is in not offering religious claims as scientific claims. I don't think a naturalistic account of the mind-body problem has been offered, and I am frankly doubtful as to whether one could be offered. But I don't think that saying we are made in the image of God is a scientific answer or adequate at this level. It may well be that we are made in the image of God, but this is simply not a scientific question. So yes I do think we can have different Magisteria but I qualify this claim quite heavily.

4. What do you consider to be your own most important contribution(s) to theorizing about science and religion?

I think I have been quite helpful in disentangling some of the major issues on the borderline between science and religion. I think in particular my 2001 book *Can a Darwinian be a Christian? The Relationship between Science and Religion* did good work in uncovering some of the issues that need to be tackled at this level. I think some of my other work in the history of science has also been useful here. I think particularly of my treatment of Darwinism and religion in my 1979 book *The Darwinian Revolution: Science Red in Tooth and Claw*, and more recently my book in 2005 *The Evolution-Creation Struggle*.

I would like to think that my most recent book on this topic in 2010, *Science and Spirituality: Making Room for Faith in the Age of Science* is an important forward-looking contribution. It is there that I developed in some detail points I was making earlier about the metaphorical nature of science and how this opens up for religious discourse. The point is that a lot of people (not just Gould) think that science and religion can co-exist—they talk in different domains. It is the position known as "neo-orthodoxy." What I do is show just why this is so, namely that because of its metaphorical nature, science doesn't speak to some questions and religion can legitimately step in. Whether religion can in fact do the job is another question. In other words, to use a term that many use critically but this I wear with pride, I think here I have made a major contribution to "accommodationism." I want to qualify the term "major." I do think my contribution is major, but it doesn't point to my brilliance. I am just using standard philosophical knowledge about the nature of theories. The point is that I have got the training. Very few working philosophically on the science-religion relationship have such training, and it shows.

I'm now working on a book on nonbelief, *Atheism: Everything You Need to Know*, and although this is basically a survey I think it is a striking statement for nonbelief. The same time I hope as always I'm able to show sympathetic understanding of the reasons for religious belief. This is a major reason why the New Atheists loathe and detest me—see what they say about me on their blogs. (Start with Chicago atheist-biologist Jerry Coyne's *Why Evolution Is True*.) Once again I think it goes back to my Quaker childhood. I hate injustice. I don't hate religion, except when it leads to injustice. I hate the aspects of the Catholic Church that led to so much child abuse and I hate even more the way the hierarchy has striven to hide things up. I don't hate someone for thinking that the bread and wine turn into the body and blood of Jesus, although I think it's pretty daft.

I think of myself as much as a historian of ideas as a straight philosopher. I think this has led me to some interesting and fruitful work. My big book, *Monad to Man: The Concept of Progress in Evolutionary Biology* has some good insights about the overall history of evolutionary theorizing—actually not just "good" insights, but really significant insights that completely rewrite the history of evolutionary theorizing (although I am not sure that anyone had really done this before except at the text book one-fact-follows-another level)—as well as about the nature of values in science. My *Evolution-Creation Struggle* is interesting for picking up on apocalyptic themes in contemporary thought. One of the things that it does, and really makes humanists and fellow travellers cross, is to suggest that thought patterns are shared by believers at both ends of the science-religion spectrum. I am hoping soon to try out the same insights on the global-warming debate. It seems to me that someone who asks us to repent our sins about carbon fuel use and go and live in yurts in North Dakota and live on raw vegetables is into religion as much as someone who argues that God gave the human race the world in Genesis One and Two and hence "what me, worry?"

My most recent book is *The Gaia Hypothesis: Science on a Pagan Planet*. It has led me to look at the Pagans, past and present. I have also looked in detail at the work and influence of Rudolf Steiner, the German clairvoyant of a hundred years ago, founder of "anthroposophy" and a major influence today through his starting the Waldorf school system. In respects he was a real fruitcake—Atlantis, two Christ figures, reincarnation, and a lot more. He was also through his followers a major although hitherto-unknown influence on Rachel Carson and through the Nobel Prize winning novelist William Golding on the author of the Gaia hypothesis, English inventor James Lovelock.

Finally I would add to a more practical level I think my contributions to the evolution-Creationism debate have been important in the courtroom, on the podium, and in print. I think particularly of my collection in 2008 *But Is It Science? The Important Question in the Evolution-Creationism Controversy*, as well as in later works including *Debating Design: Darwin to Dawkins*, a collection I put together with the Intelligent Design Theorist William Dembski.

5. What are the most important open questions, problems, or challenges confronting the relationship between science and religion, and what are the prospects for progress?

I think there's a lot of work which needs doing. I think to this date in philosophy the work has been pretty crude. I was a good friend of Steve Gould, but his work on Magisteria was embarrassing. I am ecumenical about this. Richard Dawkins on the ontological argument or God's simplicity is farcical. The same is true of the late Arthur Peacocke on the nature of causation. And I don't have much more time for Bob Russell on guided mutations at the quantum level. It is a fallacy when football players push products outside their field—"Hi, my name's Dick Butkus. Plugging holes is my job. Plugging holes is Prestone's job too." It is equally a fallacy when scientists think that their expertise in physics or chemistry or biology makes them qualified to do philosophy, without any training. I am not sure if Plato allows the Form of a "tin ear"—it may be one of those things like dirt and hair that don't make the cut—but if he does for its exemplification look no further than Jerry Coyne on free will.

I contrast this situation here with the work which has been done on the history of the science-religion relationship, for instance by Ronald Numbers, Edward Larson, and Peter Harrison. My feeling is that we need to start digging back now much more carefully than we have done previously, into the work of major theologians and philosophers and others. Although the present-day, science-religion controversy obviously has a contemporary flavor to it, I suspect that a lot of the issues with which we are dealing are not that new, and that people like St. Augustine and St. Thomas were wrestling with these and related questions. I would like to see a lot more in-the-trenches work done at this level. For myself I've got work to do on the whole question of the evolution of humankind. I touched above on the difficulties of reconciling the Darwinian views of randomness with the Christian insistence on the significance of human beings. I don't really think we've got on top of this yet, although as you see I have my own pet theory about multiverses. I want to spend at least some time working on this, although no doubt other issues will come to the fore as well. I very much hope so!

24

Robert John Russell

Robert John Russell is the Founder and Director of the Center for Theology and the Natural Sciences (CTNS), and the Ian G. Barbour Professor of Theology and Science in Residence at the Graduate Theological Union (GTU). He is the author of *Time in Eternity: Pannenberg, Physics, and Eschatology in Creative Mutual Interaction* (2012) and *Cosmology from Alpha to Omega: Towards the Mutual Creative Interaction of Theology and Science* (2008). He has also co-edited a multi-volume series of books focusing on scientific perspectives of divine action through an international research conference program co-sponsored by CTNS and the Vatican Observatory. His current research topics include: resurrection, eschatology and scientific cosmology; quantum mechanics, biological evolution, and divine action; evolution, theodicy, and Christology; philosophical assumptions in contemporary scientific cosmology and their theological roots; and time and eternity from a Trinitarian perspective in relation to time in physics.

---

1. What initially drew you to theorizing about science and religion?

The sudden death of my dad when I was twelve, and my minister's advice to me upon asking whether I'd see dad in heaven: "Because of science, we don't believe in heaven anymore." That day, in my heart, I said "no." Still my life has, in a sense, been a venture in forming an intellectually credible response, one which supports the "no," to what my minister said.

2. Do you think science and religion are compatible when it comes to understanding cosmology (the origin of the universe), biology (the origin of life and of the human species), ethics, and/or the human mind (minds, brains, souls, and free will)?

Yes, and the wealth of scholarly literature in theology and science demonstrates it, including my own numerous writings.

So, for example, I have argued that the beginning of the universe, "t=0," according to standard Big Bang cosmology, is a positive factor in an assertion that the sheer existence of the universe supports the essence of the doctrine of creation out of nothing (*ex nihilo*): a universe with a beginning is clearly contingent and need not exist. I describe t=0 as a "character witness" in a court of law, supporting the existence of God without actual, direct evidence for God's existence (which, of course, we could never have as classical theism clearly argues). Of course recent developments in inflationary Big Bang, and such multiverse theories as inflationary cosmology and superstring multiverses, undercut the likelihood that our universe had a beginning in time ("t=0") but they do nothing to challenge the ontological fact—and mystery—that the universe exists *per se* regardless of whatever form its past existence takes (infinite past, finite past, etc.).

Regarding the origins of the human species, many authors have supported what is generically called "theistic evolution," that evolution is how God creates the diversity of species on earth (and perhaps throughout the universe). Voices include Pope John Paul II, Ian Barbour, Teilhard de Chardin, Denis Edwards, John Haught, Ernan McMullin, Jürgen Moltmann, Nancey Murphy, Wolfhart Pannenberg, Arthur Peacocke, Ted Peters & Martinez Hewlett, and Karl Rahner to mention only a few. Note: none of these would have any interest in "Intelligent Design," let alone "Creation Science." Finally, most scholars in theology and science support an integrated view of the human person, in which "mind" and "brain" are not separate entities (as in Plato and Descartes) but two irreducible aspects of the human person (e.g., "dual-aspect monism"). Of course reductionists like the Churchlands dispute this, but scholars like Ian Barbour, Nancey Murphy, and Arthur Peacocke, have made, in my view, quite convincing arguments against this sort of reductionism.

3. Some theorists maintain that science and religion occupy non-overlapping magisteria—i.e., that science and religion each have a legitimate magisterium, or domain of teaching authority, and these two domains do not overlap. Do you agree?

No, this is an entirely outmoded view, whose intellectual roots lie in the Enlightenment/18th century thought of Kant, ideas which were recently reclaimed in a rather superficial way, by Stephen Gould ("NOMA").

4. What do you consider to be your own most important contribution(s) to theorizing about science and religion?

That theology can contribute research directions, suggestions, and questions to science, making theology and science a true interaction (or what I call "Creative Mutual Interaction," CMI, especially paths 6-8 from theology to science).

5. What are the most important open questions, problems, or challenges confronting the relationship between science and religion, and what are the prospects for progress?

Clearly scientific cosmology, with the prediction of an ever expanding and freezing universe, severely challenges Christian eschatology, with its hopes for a New Creation of endless, eternal life. Included here is the challenge of what bodily resurrection means and what science might or might not say about it.

25

John Searle

John Searle is a world-renowned American philosopher, widely recognized for his contributions to the philosophy of language, philosophy of mind, and social philosophy. He is currently Slusser Professor of Philosophy at the University of California, Berkeley, and is the recipient of numerous awards, including the Jean Nicod Prize in 2000, the National Humanities Medal in 2004, and the Mind and Brain Prize in 2006. Searle is the author of dozens of scholarly articles and over a dozen books, including *Speech Acts: An Essay in the Philosophy of Language* (1969), *Expression and Meaning: Studies in the Theory of Speech Acts* (1979), *Intentionality: An Essay in the Philosophy of Mind* (1983), *Minds, Brains, and Science* (1984), *The Rediscovery of Mind* (1992), *The Construction of Social Reality* (1995), *The Mystery of Consciousness* (1997), *Mind, Language and Society: Philosophy in the Real World* (1998), *Rationality in Action* (2001), *Freedom and Neurobiology* (2004), *Mind: A Brief Introduction* (2004), *Philosophy in a New Century: Selected Essays* (2008), and *Making the Social World: The Structure of Human Civilization* (2010).

---

1. What initially drew you to theorizing about science and religion?

I have never theorized about science and religion. These are names of domains of activity and not the names of theories. "Science" is the name of a set of rational activities designed to try to get at the truth about how nature works. "Religion" is not in that way the well-defined set of activities because there are all sorts of different sorts of religions.

2. Do you think science and religion are compatible when it comes to understanding cosmology (the origin of the universe), biology (the origin of life and of the human species), ethics, and/or the human mind (minds, brains, souls, and free will)?

It depends on which science and which religion you are talking about. There are religions that deny evolution and they are obviously false. The aim of science is specifically to state the truth. To the extent that any religion denies the truth it is stating something false.

3. Some theorists maintain that science and religion occupy non-overlapping magisteria—i.e., that science and religion each have a legitimate magisterium, or domain of teaching authority, and these two domains do not overlap. Do you agree?

I don't think this view can be stated coherently. Anything you can study systematically is legitimately the subject matter for science. If you get a well-established result and some "religion" denies it, you know that the religion made a mistake. If God really exists that would be a scientific fact like any other. There could not be a special domain of "religion" because any domain that admits of systematic study can be a subject matter of a science.

4. What do you consider to be your own most important contribution(s) to theorizing about science and religion?

As far as I know I have made no contribution whatever to theorizing about science and religion, because I consider it a fruitless enterprise. The two concepts are too vague to be worth theorizing about.

5. What are the most important open questions, problems, or challenges confronting the relationship between science and religion, and what are the prospects for progress?

As far as I know there are no interesting philosophical questions in this domain. There are some interesting psychological and sociological questions: such as, how is it that traditional dogmatic religions survive when their views are too idiotic to be worth even discussing. My guess is the traditional religions satisfied deep psychological needs, and the fact that their views are often idiotic is not a source of weakness but a source of strength.

26

Michael Shermer

Michael Shermer is the Founding Publisher of *Skeptic* magazine and editor of Skeptic. com, a monthly columnist for *Scientific American*, and an Adjunct Professor at Claremont Graduate University and Chapman University. Dr. Shermer's latest book is *The Believing Brain: From Ghosts and Gods to Politics and Conspiracies—How We Construct Beliefs and Reinforce Them as Truths*. His last book was *The Mind of the Market*, on evolutionary economics. He also wrote *Why Darwin Matters: Evolution and the Case Against Intelligent Design*, and he is the author of *The Science of Good and Evil* and of *Why People Believe Weird Things*. Dr. Shermer received his B.A. in psychology from Pepperdine University, M.A. in experimental psychology from California State University, Fullerton, and his Ph.D. in the history of science from Claremont Graduate University (1991). He was a college professor for twenty years, and since his creation of *Skeptic* magazine he has appeared on such shows as *The Colbert Report*, *20/20*, *Dateline*, *Charlie Rose*, and *Larry King Live*. Dr. Shermer was the co-host and co-producer of the 13-hour Family Channel television series, *Exploring the Unknown*.

---

1. What initially drew you to theorizing about science and religion?

I have been interested in the subject since my late teen years in high school when I became a born-again Christian at the urging of my best friend, who was a recent convert himself. Now, many people pass quickly through such phases, but I took my conversion seriously and I became profoundly religious, fully embracing the belief that Jesus suffered wretchedly and died, not just for humanity, but for me personally. It felt good. It seemed real. And for the next seven years I walked the talk. Literally. I went door-to-door and person-to-person, witnessing for God and evangelizing for Christianity. I became a "bible thumper," as one of my friends called me, a "Jesus freak" in the words of a sibling. A little religion is one thing, but when it is all one talks about it can become awkward and uncomfortable for family and friends who don't share your faith passion. When I got to college, however, my professors did not just blow me off but instead engaged me in spirited conversation and debate about the merits of the specific tenets of my faith. This forced me to read deeply not just in the theological apologetics of Christian theologians, but in religious philosophy and the science and religion debate as well. I matriculated at Pepperdine University, a Church of Christ institution that mandated chapel attendance twice a week, along with a curriculum that included courses in the Old and New Testaments,

the life of Jesus, and the writings of C. S. Lewis. Although all this theological training would come in handy years later in my public debates
on God, religion, and science, at the time I studied it because I believed
it, and I believed it because I unquestioningly accepted God's existence
as real, along with the resurrection of Jesus, and all the other tenets of
the faith. Later, when I lost my faith and became an atheist, and turned
my scholarly attention to the scientific study of religion, I was better
able to appreciate the mind of the believer because I was once one—a
True Believer—and that has enabled me to better mind read what is
behind the beliefs and arguments of the faithful today.

**2. Do you think science and religion are compatible when it comes
to understanding cosmology (the origin of the universe), biology
(the origin of life and of the human species), ethics, and/or the
human mind (minds, brains, souls, and free will)?**

For me question numbers 2 and 3 are the same. I answer them both
below.

**3. Some theorists maintain that science and religion occupy non-
overlapping magisteria—i.e., that science and religion each have a
legitimate magisterium, or domain of teaching authority, and these
two domains do not overlap. Do you agree?**

On a superficial and social level, Stephen Jay Gould's notion of
NOMA—Non-Overlapping Magisteria—works well to keep the peace
between science and religion, and on one level I agree with *The Simp-
sons* take on the matter in the episode in which Gould makes a guest
appearance and a judge imposed a restraining order on religion that it
was to stay 500 yards away from science at all times. When religion is
doing things like staffing soup kitchens and bringing relief to people
devastated by hurricanes, tornados, and earthquakes, there is obviously
no conflict with science because scientists—when they are practicing
science—do not do such things as part of their job description. Of
course, scientists may be generous and helpful and charitable and do-
nate their time to such worthy causes as helping the poor, but when they
do so they are not acting as scientists per se.

When you push the matter deeper, however, science and religion are
ultimately incompatible because their magisterial do overlap. Science
operates in the natural, not the supernatural. In fact, I go so far as to
argue that there is no such thing as the supernatural or the paranormal. There is just the natural, the normal, and mysteries we have yet
to explain by natural causes. Invoking such words as "supernatural"
and "paranormal" just provides a linguistic place-holder until we find
natural and normal causes, or we do not find them and discontinue the

search out of lack of interest. This is what normally happens in science. Mysteries once thought to be supernatural or paranormal happenings—such as astronomical or meteorological events—are incorporated into science once their causes are understood. For example, when cosmologists reference "dark energy" and "dark matter" in reference to the so-called "missing mass" needed to explain the structure and motion of galaxies and galaxy clusters, they do not intend these descriptors to be causal explanations. Dark energy and dark matter are merely cognitive conveniences until the actual sources of the energy and matter are discovered. When theists, creationists, and Intelligent Design theorists invoke miracles and acts of creation *ex nihilo*, that is the end of the search for them. For scientists, the identification of such mysteries is only the beginning. Science picks up where theology leaves off. When a theist says "and then a miracle happens," as wittily portrayed in my favorite Sydney Harris cartoon of the two mathematicians at the chalkboard with the invocation tucked in the middle of a string of equations, I quote from the cartoon's caption: "I think you need to be more explicit here in step two."

To our bronze-age ancestors who created the great monotheistic religions, the ability to create the world and life was God-like. Once we know the technology of creation, however, the supernatural becomes the natural. Thus, the only God that science could discover would be a natural being, an entity that exists in space and time and is constrained by the laws of nature. A supernatural God who exists outside of space and time is not knowable to science because He is not part of the natural world, and therefore science cannot know God. Q.E.D.

4. What do you consider to be your own most important contribution(s) to theorizing about science and religion?
Two:

1. Why People Believe in God. In my book, *How We Believe*, I reported the results of a study conducted by the U.C. Berkeley social scientist Frank J. Sulloway on why people believe in God and why people believe other people believe in God. We discovered another form of what is called the *attribution bias*, or *the tendency to attribute different causes for our own beliefs and actions than that of others*. There are several types of attribution bias. There is a *situational attribution bias* in which we identify the cause of someone's belief or behavior in the environment ("her success is a result of luck, circumstance, and having connections") and a *dispositional attribution bias*, in which we identify the cause of someone's belief or behavior in the person as an enduring personal trait ("her

success is due to her intelligence, creativity, and hard work"). And, thanks to the self-serving bias, we naturally attribute our own success to a positive disposition ("I am hard working, intelligent, and creative") and we attribute others' success to a lucky situation ("he is successful because of circumstance and family connections"). In our study Sulloway and I polled 10,000 random Americans in which we directly asked them in an essay question why they believe in God and why they think others believe in God. The top two reasons that people gave for why they believe in God were "the good design of the universe" and "the experience of God in everyday life." Interestingly, and tellingly, when asked why they think other people believe in God, these two answers dropped to sixth and third place, respectively, while the two most common reasons given were that belief is "comforting" and "fear of death." These answers reveal a type of belief attribution bias in which there is a sharp distinction between an *intellectual attribution bias* in which people consider their own beliefs as being rationally motivated, and an *emotional attribution bias* in which people see the beliefs of others as being emotionally driven.

2.   A gambit that I call Shermer's Last Law: *Any sufficiently advanced Extra-Terrestrial Intelligence is indistinguishable from God.* Most theists believe that God created the universe and everything in it, including stars, planets, and life. My question is this: how could we distinguish an omnipotent and omniscient God or Intelligent Designer (ID) from an extremely powerful and really smart Extra-Terrestrial Intelligence (ETI)? My gambit (ET = ID = God) arises from an integration of evolutionary theory, Intelligent Design creationism, and the SETI (Search for Extra-Terrestrial Intelligence) program, and can be derived from the following observations and deductions.

Observation I. Biological evolution is glacially slow compared to technological evolution. The reason is that biological evolution is Darwinian and requires generations of differential reproductive success, whereas technological evolution is Lamarckian and can be implemented within a single generation.

Observation II. The cosmos is very big and space is very empty, so the probability of making contact with an ETI is remote. By example, the speed of our most distant spacecraft, *Voyager I*, relative to the sun is 17.246 kilometers per second, or 38,578 miles per hour. If *Voyager I* was heading toward the closest star system to us (which it isn't)—the Alpha Centauri system at 4.3 light years away—it would take an almost unfathomable 74,912 years to get to there.

Deduction I. The probability of making contact with an ETI who is only slightly more advanced than us is virtually nil. Any ETIs we would encounter will either be way behind us (in which case we could only encounter them by landing on their planet) or way ahead of us (either through telecommunications or by landing on our planet). How far ahead of us is an ETI likely to be?

Observation III. Science and technology have changed our world more in the past century than it changed in the previous hundred centuries—it took 10,000 years to get from the cart to the airplane, but only 66 years to get from powered flight to a lunar landing. Moore's Law of computer power doubling every eighteen months continues unabated and is now down to about a year. Computer scientists calculate that there have been thirty-two doublings since World War II, and that as early as 2030 we may encounter the Singularity—the point at which total computational power will rise to levels that are so far beyond anything that we can imagine that they will appear near infinite and thus, relatively speaking, be indistinguishable from omniscience. When this happens the world will change more in a decade than it did in the previous thousand decades.

Deduction II. Extrapolate these trend lines out tens of thousands, hundreds of thousands, or even millions of years—mere eye blinks on an evolutionary time scale—and we arrive at a realistic estimate of how far advanced an ETI will be. Consider something as relatively simple as DNA. We can already engineer genes after only 50 years of genetic science. An ETI that was 50,000 years ahead of us would surely be able to construct entire genomes, cells, multi-cellular life, and complex ecosystems. (At the time of this writing the geneticist J. Craig Venter produced the first artificial genome and constructed a synthetic bacteria that was chemically controlled by the artificial genome.) The design of life is, after all, just a technical problem in molecular manipulation. To our not-so-distant descendents, or to an ETI we might encounter, the ability to create life will be simply a matter of technological skill.

Deduction III. If today we can engineer genes, clone mammals, and manipulate stem cells with science and technologies developed in only the last half century, think of what an ETI could do with 50,000 years of equivalent powers of progress in science and technology. For an ETI who is a million years more advanced than we are, engineering the creation of planets and stars may be entirely possible. And if universes are created out of collapsing black holes—which some cosmologists think is probable—it is not inconceivable that a sufficiently advanced

ETI could even create a universe by triggering the collapse of star into a black hole.

What would we call an intelligent being capable of engineering life, planets, stars, and even universes? If we knew the underlying science and technology used to do the engineering, we would call it an Extra-Terrestrial Intelligence; if we did not know the underlying science and technology, we would call it God.

5. What are the most important open questions, problems, or challenges confronting the relationship between science and religion, and what are the prospects for progress?

Now it is time to step out of our evolutionary heritage and our historical traditions and embrace science as the best tool ever devised for explaining how the world works, and to work together to create a social and political world that embraces moral principles and yet allows for natural human diversity to flourish. Religion cannot get us there because it has no systematic methods of explanation of the natural world, and no means of conflict resolution on moral issues when members of competing sects hold absolute beliefs that are mutually exclusive. Flawed as they may be, science and the secular Enlightenment values expressed in Western democracies are our best hope for survival. The arc of the moral universe is long but it bends toward justice, and most of the work has been done since the Enlightenment and as a result of secular science not religious faith. The next area of progress will be found in developing a science of morality and human flourishing.

27

Victor J. Stenger

Victor J. Stenger is Emeritus Professor of Physics and Astronomy at the University of Hawaii and Adjunct Professor of Philosophy at the University of Colorado. Dr. Stenger's research career spanned the period of great progress in elementary particle physics that ultimately led to the current standard model. He participated in experiments that helped establish the properties of strange particles, charmed quarks, gluons, and neutrinos. He also helped pioneer the emerging fields of very high-energy gamma ray and neutrino astronomy. In his last project before retiring, Dr. Stenger collaborated on the underground experiment in Japan that showed for the first time that the neutrino has mass. The Japanese leader of the project shared the 2002 Nobel Prize in physics for that work. Dr. Stenger is the author of twelve books including the 2007 *NY Times* bestseller *God: The Failed Hypothesis—How Science Shows That God Does Not Exist*. His other books include *Not by Design: The Origin of the Universe* (1988), *Physics and Psychics: The Search for a World Beyond the Senses* (1990), *The Unconscious Quantum: Metaphysics in Modern Physics and Cosmology* (1995), *Has Science Found God? The Latest Results in the Search for Purpose in the Universe* (2003), *The New Atheism: Taking a Stand for Science and Reason* (2009), *The Fallacy of Fine-Tuning: How the Universe is Not Designed for Humanity* (2011), *God and the Folly of Faith: The Incompatibility of Science and Religion* (2012), *God and the Atom: From Democritus to the Higgs Boson* (2013), and *God and the Multiverse: Humanity's Expanding View of the Cosmos* (in press).

---

1. What initially drew you to theorizing about science and religion?

Theorizing about anything in the absence of empirical data is a largely useless activity. A logical deduction tells you nothing that is not already embedded in its premises. For the deduction to have value, those premises must be based on empirical data. All my theorizing about science and religion refers to well-established, objective observations.

I believe strongly in the integrity of the scientific method. In the 1980s, I became aware of how science was being misused to claim evidence for special powers of the mind, such as extrasensory perception (ESP). I looked at the history and saw that the scientific study of so-called *psychic* phenomena began in the late nineteenth century, coinciding with the rising public interest in spiritualism at that time.

Some of this research was being conducted by reputable scientists, notably physicists William Crookes and Oliver Lodge. I discovered that each was strongly motivated by a personal tragedy that led him to desperately seek evidence for life after death. Psychic phenomena might substantiate the existence of an immortal soul.

At the time, it was not unreasonable for a scientist to consider the possibility of psychic powers. As Lodge, who had helped develop radio, put it: "If wireless telegraphy is possible, why not wireless telepathy?" He and Crookes were also impressed by the many remarkable feats that were being demonstrated by famous "psychic mediums" of the day. They each conducted tests of mediums using what they genuinely thought were carefully controlled scientific experiments.

Unfortunately, their desire to believe led Crookes and Lodge to relax their standards and to fail to apply the strict criteria they might have for a less emotionally charged subject. And so, instead of conducting experiments in their own labs under careful conditions designed by themselves, they carried out the experiments on the psychics' own turf with the experimental subjects actually calling the shots. This usually involved observing séances in darkened rooms where they gullibly fell for magician's tricks that were pretty standard in the professional illusionist business but were unfamiliar to the trusting scientists who were used to nature not lying to them.

By the turn of the century, skeptics had successfully debunked all the most notable claims, and psychic research, which came to be known as parapsychology, fell into disrepute. It was revived in the 1930s by a botanist at Duke University, Joseph Banks Rhine, who at first appeared to bring a new level of scientific integrity to psychic studies. He coined the term "ESP" (Extra-Sensory Perception) in a very popular book, and his experiments were conducted under what on the surface were well-controlled conditions.

The media picked up on Rhine's claims of success and science fiction writers began to include characters with psychic powers in their tales, giving the idea at least an aura of scientific respectability. Years later we would still have Mr. Spock reading minds in the Star Trek space sagas. But that had no more basis in serious science than warp drives, tracking beams, or crewmembers being able to breathe the air on every planet they beamed down to.

While there were some cases of proved cheating in Rhine's lab, he is reputed to have been scrupulously honest. However, he too was fervently motivated by religion that, as with Crookes and Lodge, affected his critical capabilities. Rhine always found excuses for negative results rather than simply accepting them as facts and when scientific journals starting rejecting his papers he started his own "peer-reviewed" journal. Today Rhine's claims are not taken seriously by mainstream science.

When I saw that not only parapsychology, but other fields such as medical research and psychology had journals with low publication standards I started speaking out against them. You still often read that a claimed effect is "statistically significant" if it has confidence level of 5 percent, which is the standard for these journals. What this means is that if you were to repeat the same experiment many times under the exact same conditions, then you would get the observed effect or a greater one no more than one out of 20 times on average simply by chance. This implies that if 100 experiments are conducted in a variety of fields, five of them will be published having effects that are taken to be significant but are really just statistical fluctuations. And, it's worse than that. A good number of the 95 other experiments that saw "no effect" are likely not to be published, since only positive results tend to be published. In fact, it would be a reasonable conclusion that most published claims of new phenomena at the 5 percent level are simply wrong.

By the late 1990s I grew bored of shooting arrows into the dead horse of psychic claims and found a new animal to target. I discovered that many Christian apologists were arguing that science was finding evidence for God in cosmology. Again I saw science being misused so I took action.

2. Do you think science and religion are compatible when it comes to understanding cosmology (the origin of the universe), biology (the origin of life and of the human species), ethics, and/or the human mind (minds, brains, souls, and free will)?

Science and religion are fundamentally incompatible because of the different assumptions they make about how humans gain knowledge of the world. Religious belief is based on the notion that a world exists that we can access purely mentally outside the material world of our senses. A God who communicates directly with humans by means of revelation conflicts with the fact that no scientifically verifiable new information has ever been transmitted while many wrong and harmful doctrines have been asserted by this means. Furthermore, physical evidence now conclusively demonstrates that some of the most important biblical narratives, such as the Exodus, never took place.

Physical and historical evidence might have been found for the miraculous events and the important narratives of the scriptures. For example, Roman records might have been found for an earthquake in Judea at the time of a certain crucifixion ordered by Pontius Pilate. Noah's Ark might have been discovered. The Shroud of Turin might have contained genetic material with no Y-chromosomes. Since the image is that of a man with a beard, this would confirm he was born of a virgin. Or, the genetic material might contain a novel form of coding molecule not found in any other living organism. This would have proven an alien (if not divine) origin of the enshrouded being.

In fact, there isn't a shred of independent evidence that Jesus Christ is a historical figure.

If God exists, then miracles that violate scientific principle should be seen. For example, prayers should be answered; an arm or a leg should be regenerated through faith healing. This does not happen. Let us now see how in each of the above subjects scientific understanding differs irreconcilably from religious beliefs.

Cosmology

I know of no major religion except Buddhism that does not believe that the universe was supernaturally created a finite time in the past. The consensus view of scientific cosmology holds that natural origin of our universe does not violate any principles of existing, well-established physics and cosmology. Furthermore, modern cosmology suggests that our universe is one of an *eternal multiverse* containing an unlimited number of universes. That multiverse had no beginning, nor will it end, and so no supernatural creation *ex-nihilo* was necessary.

A cosmic God who fine-tuned the laws and constants of physics for life, in particular human life, fails to agree with the fact that the universe is not congenial to human life, being from the human perspective tremendously wasteful of time, space, and matter. It also fails to agree with the fact that the universe is mostly composed of particles in random motion, with complex structures such as galaxies forming less than 4% of the total mass of the universe.

If humanity were a special creation of God, the universe should be congenial to human life. Humans might have been able to move from planet to planet, just as easily as they now move from continent to continent, and be able to survive on every planet—even in space—without artificial life support.

In short, the picture that modern physics and cosmology draws of the universe is in complete contradiction to that presented in the scriptures of most religions, especially Judaism, Christianity, and Islam.

Biology

The Catholic Church and many moderate Protestant denominations claim they accept biological evolution. However, when you read what they actually say about the subject you find that what they believe in is not Darwinian evolution as understood by science. Rather it is God-guided evolution that is simply another form of "intelligent design."

Ethics

Considerable scholarship now exists describing how humans, and even many animals, evolved morals and ethics naturally in order to live together in societies. Religions universally claim morals and ethics arose from supernatural origins. This simply disagrees with the historical facts. For example, the Golden Rule existed in many cultures a thousand years before The Sermon On the Mount.

It is now clear that social animals evolved ethics even before humans. People have further developed ethics and culture by trial and error. Every improvement in ethics was resisted by religious conservatives at the time.

If morality came from God, you would expect that natural events should follow some moral law, rather than such events following morally neutral physical laws. For example, lightning might strike only the wicked, people who behave badly might fall sick more often, nuns would always survive plane crashes.

Likewise, believers should be expected to have a higher moral sense than nonbelievers and other measurably superior qualities. For example, the jails might be filled with atheists while all believers live happy, prosperous, contented lives surrounded by loving families and pets. But the opposite is true.

Human Mind

All the evidence points to the mind and consciousness being a product of the purely material brain. A God who has given humans immortal souls fails to agree with the empirical facts that human thoughts, memories, and personalities are governed by physical processes in the brain, which dissolves upon death. And, as shown earlier, no nonphysical or extra-physical powers of "mind" can be found and no evidence exists for an afterlife

Science might have uncovered convincing evidence for an afterlife. For example, a person who had been declared dead by every means known to science might return to life with detailed stories of an afterlife that were later verified. She might meet Jimmy Hoffa who tells her where to find his body.

Similarly, any claim of a revelation obtained during a mystical trance could contain scientifically verifiable information that the subject could not possibly have known. Yet none of this has happened.

3. Some theorists maintain that science and religion occupy non-overlapping magisteria—i.e., that science and religion each have a legitimate magisterium, or domain of teaching authority, and these two domains do not overlap. Do you agree?

No I do not. This was the good-intentioned attempt by the late Stephen Jay Gould, a professed atheist, to reduce the conflict between science and religion. Science, Gould wrote, is concerned with describing the "outer" world of our senses, while religion deals with the "inner" world of morality and meaning.

Many scientists—believers and nonbelievers—have adopted the NOMA position. Believing scientists compartmentalize their thinking by not incorporating into their religious thinking the *doubt-everything* position they were trained to take in their professions.

A prime example is geneticist, Francis Collins, who directed the Human Genome Project and at this writing directs the National Institutes of Health. His 2006 book *The Language of God: A Scientist Presents Evidence for Belief*, was a bestseller. However, his so-called evidence was not, as you might have thought from the title, based on his deep knowledge of DNA. Rather it was based on his own inner feeling that the world is a moral place and only God could have made it that way. Nowhere does Collins come close to applying to this notion the critical skills exhibited in his outstanding scientific career.

Unlike Descartes, Newton, Kepler and many of the great founders of the scientific revolution (Galileo is a prominent exception), modern-day believing scientists such as Collins do not incorporate God into their science. This even includes those scientists who happen to also be members of holy orders, such as the Belgian Catholic priest Georges-Henri Lemaître who proposed the big bang in 1927 but urged Pope Pius XII not to claim it as infallible proof that God exists.

Most nonbelieving scientists just want to do their research and stay out of any fights over religion. That makes the NOMA approach appealing because it allows these scientists to not worry much about what religion is or how it affects our social and political world. In my view, though, these scientists are shirking their responsibility by conceding the realms of morality and public policy to the irrationality and brutality of faith.

4. What do you consider to be your own most important contribution(s) to theorizing about science and religion?

I believe I was the first to argue, in my 2007 book *God—The Failed Hypothesis*, that the absence of evidence for any God who plays an important role in the universe proves beyond a reasonable doubt that such a god does not exist.

It is inarguable that science has not yet found evidence for a god or the supernatural. If it had, it would be in the textbooks. Still, you will often hear: "Absence of evidence is not evidence of absence." Not always. When the evidence that is absent is evidence that should be there, then that can be taken as evidence of absence.

Consider the hypothesis that elephants roam Yellowstone Park. If that were the case, then a tourist or ranger should have been spotted one by now. Or, other evidence such as droppings and crushed bushes would certainly be found. Since none of this has occurred, we can conclude beyond a reasonable doubt that elephants do not roam Yellowstone Park.

Incidentally, note that the often-heard statement that you cannot prove a negative is simply wrong, including the negative that there is no God. And so it is with the hypothesis of the existence of the God worshipped by Jews, Christians, and Muslims and others among the major religions of the world. Religious apologists and even some atheistic scientists have contended that God is a "spirit" and so science can say nothing about him. However, this particular hypothesized God plays such an active part in the universe that evidence of his actions should be observable.

Earlier I discussed several examples of phenomena that should be seen if such a God exists. The fact that they are not is proof beyond a reasonable doubt that this God does not exist. Note that this proof does not apply to all conceivable gods, such as the impersonal deist god who creates the universe but does not intervene any further. While such a god is not ruled out, we have no reason to pray to it or worship it, so it might as well not exist.

In addition, I have provided unique counter examples to the questions often raised by believers that claim to show the need for *some* kind of god.

How can something come from nothing?

Let me restate this question as follows: How can matter come from non-matter?

The universe has mass, which is a measure of the amount of matter in a body. Since mass and rest energy are equivalent, it would seem that the law of conservation of energy must have been broken to create the matter of the universe out of "nothing."

However, when the negative potential energy of attractive gravity is included, the total energy of the universe is zero, give or take quantum uncertainties. So, the law of conservation of energy was not broken for the universe to appear from an earlier state of zero energy and zero matter. This leads to the next question:

Where do the laws of physics come from?

The "law" of conservation of energy follows from the fact that there is no special moment in time. It was not handed down in a stone tablet by God. It's more necessary than God.

Physicists in the twentieth century discovered a set of principles I call *metalaws* that are required to be present in all physics models. In order to describe the universe objectively, physicists must formulate their models so that they describe observations in ways that are independent of the point-of-view of particular observers, what I have dubbed *point-of-view invariance*. This gives the model-builders no choice but to include metalaws—the great principles of conservation of energy, linear momentum, angular momentum, and electric charge.

I have shown that point-of-view invariance also leads to classical physics, including Newton's laws of mechanics and gravity, Maxwell's equations of electromagnetism, and Einstein's theory of special relativity. Much if not all of general relativity and quantum mechanics, including the Heisenberg uncertainty principle, also follow.

I have also shown that the parameters of physics that are supposedly "fine-tuned" for life are consistent with known physics and capable of producing some form of life. Apologists come back with, "Where did physics come from?" My answer: physics came from physicists formulating models to describe observations and these models must include metalaws that constitute the basic laws of physics. The metalaws do not set all the parameters of physics. Many are determined by accident. However, the values of the parameters in the models that successfully describe all observations in our universe are within the ranges set by the metalaws.

Why is there something, rather than nothing?

This question is largely philosophical because it deals more with the meaning of words than actual physics. Clearly, no consensus exists on how to define "nothing." It may be impossible. To define "nothing" you have to give it some defining property, but, then, if it has a property it's not nothing!

Let me ask the questions another way: Why is there "being" rather than "nonbeing"? My reply: Why should nonbeing, no matter how defined, be the default state of existence rather than being? Why is some creative act needed to convert nonbeing to being? Perhaps such an act is needed to convert being into nonbeing.

If nonbeing is the natural state, then why is there God? Once theologians assert that there is a God as opposed to nonbeing, they can't turn around and demand that a cosmologist explain why there is a universe as opposed to nothing. They claim God is a necessary entity. Why can't a godless multiverse be a necessary entity?

But we can do even better than this standoff and make an argument for "something" being a more natural state than "nothing." We can provide a plausible reason based on our knowledge of existing physics.

It is commonly thought that a complex physical system can only come about by the deliberate act of an intelligent designer who must necessarily be even more complex. The chain of design then leads back to God as Aristotle's Prime Mover and Aquinas's First Cause Uncaused, the maximally complex creator of all that is.

We even do not have to rely on sophisticated scientific arguments to see from common experience alone that Aristotle and Aquinas had it backward. In nature, complexity arises out of simplicity. Consider the phase transitions observed in familiar matter. In the absence of external heat, water vapor will naturally condense into liquid water, which then will freeze into solid ice. With each transition, we move from a state of higher symmetry to one of lower symmetry—from simplicity to complexity.

Complexity is broken symmetry, and the transition from simple to complex occurs spontaneously. Simplicity begets complexity, not the other way around. The particular crystal structure that results from the liquid-water-to-ice transition is unpredictable, that is, accidental.

Physical systems move naturally from simple to complex without the need for design, intelligent or otherwise. Indeed, the fact that specific events, such as atomic transitions, are random can be taken as strong evidence against any design, intelligent or not so intelligent.

And so, how do we get something from nothing? Since no thing is more symmetric than nothing, we would expect nothing to naturally undergo a phase transition to something. As Nobel laureate Frank Wilczek put it in a *Scientific American* article back in 1980, "Nothing is unstable."

5. What are the most important open questions, problems, or challenges confronting the relationship between science and religion, and what are the prospects for progress?

1. Science needs to produce life in the laboratory. It does not have to be exactly our form of life, with DNA, etc., but just a material system with the basic characteristics of life. This will prove that a supernatural force was not needed to produce life and it can originate naturally. It may take a while.

2. Science needs to find life on another planet. This will prove that our form of life is not the only in existence. This will prove that earthly life is not special. It could happen soon.

3. Science needs to demonstrate conclusively that the thinking process is purely material. This will disprove the existence of an immaterial soul, at least as traditionally conceived. We are almost there.

4. Those religions whose traditions are based on the knowledge of

simple desert tribesman who lived thousands of years ago need to show how those traditions can be reconciled with modern science. To those ancients, Earth was the unmovable center of the universe, and this is how the Bible describes it. Today we know that the universe is vast beyond comprehension and probably contains countless other intelligent beings. Do they all require the only begotten son of God to visit them and die on their planet to atone for their sins? I don't see how Christian traditions will ever be reconciled with scientific knowledge.

5. Religious believers need to place empirical knowledge and reason ahead of baseless faith in making decisions in life. I think this will gradually happen, as religion is maintained mainly for its cultural heritage and social benefits while its supernatural elements fade away as more people develop critical thinking skills.

6. Young people are moving away from organized religion in droves. Within another generation it is likely that believers in the supernatural will be a minority in America as they are in most of Europe. However, many of these so-called "nones" currently say they are "not religious but spiritual." The meaning of the word "spiritual" must be made clear. Its general usage connotes something supernatural. We have seen that all attempts to find evidence for a world beyond matter have failed. It would be tragic if the primitive superstitions of traditional religions are simply replaced by a different system based on just another form of superstition. "Not religious but spiritual" should be replaced with "not religious but moral."

From its very beginning, religion has been a tool used by those in power to retain that power and keep the masses in line. This continues today as religious groups are manipulated to work against believers' own often unrecognized best interests in health and economic well-being in order to cast doubt on well-established scientific findings. This would not be possible except for the diametrically opposed world-views of science and religion and the illegitimate and unearned power of the latter.

I have an urgent plea to scientists and all thinking people. We need to focus our attention on one goal, which will not be reached in the lifetime of the youngest among us but has to be achieved someday if humanity is to survive. That goal is the replacement of foolish faith and its vanities with something more sublime—knowledge and understanding that is securely based on observable reality.

28

Robert Thurman

Robert Thurman is the Jey Tsong Khapa Professor of Indo-Tibetan Buddhist Studies in the Department of Religion at Columbia University, President of the Tibet House U.S., a non-profit organization dedicated to the preservation and promotion of Tibetan civilization, and President of the American Institute of Buddhist Studies, a non-profit affiliated with the Center for Buddhist Studies at Columbia University. *Time* magazine chose Thurman as one of its 25 most influential Americans in 1997, describing him as a "scholar-activist destined to convey the Dharma, the precious teachings of Siddhartha, from Asia to America," and *The New York Times* recently said Thurman "is considered the leading American expert on Tibetan Buddhism." Thurman is the author of many books on Tibet, Buddhism, art, politics, and culture, including *Wisdom and Compassion: The Sacred Art of Tibet* (with Marilyn Rhie) (1992), *The Tibetan Book of the Dead* (1994), *Essential Tibetan Buddhism* (1995), *Inner Revolution: Life, Liberty, and the Pursuit of Real Happiness* (1998), *Circling the Sacred Mountain* (with Tad Wise) (1999), *Infinite Life: Awakening the Bliss Within* (2003), *Anger* (2004), *Why the Dalai Lama Matters: His Act of Truth as the Solution for China, Tibet, and the World* (2008), and *Love Your Enemies: It Will Drive Them Crazy* (with Sharon Salzberg) (2013). Thurman was the first American to be ordained a Tibetan Buddhist monk and has been a close friend of His Holiness the 14th Dalai Lama for 50 years.

---

1. What initially drew you to theorizing about science and religion?

A feeling that I had throughout my education, St. Bernard's, Phillips Exeter, and Harvard, that the US culture was incomplete and confused. Neither science nor religion had it together in my view, the former emphasizing reason and knowledge but still convinced a priori that there is no ultimate knowing, and the latter certain there is no ultimate knowing and so one should blindly believe against reason. So I read everywhere looking for a meaningful answer and when I had finished up to Wittgenstein and Tillich etc., I turned East, and I found the path to the answers I have been satisfied with, though I have further to go of course, in the Centrist thought of Nāgārjuna and the centuries of other teachers and saints among the Mahāyāna Buddhists of India and Tibet.

2. Do you think science and religion are compatible when it comes to understanding cosmology (the origin of the universe), biology (the origin of life and of the human species), ethics, and/or the human mind (minds, brains, souls, and free will)?

Western science and religion are certainly not compatible, though scientists are more under the influence of western religions than they think. Ultimately, there is no reason why those who seek to know accurately the nature of the self and the universe should not succeed, and once they do, they invariably realize that the good that is sought by religions emerges from the true nature of reality. Thus, all the italicized concepts above (e.g., cosmology, biology, ethics, and the human mind) can be nicely fit together in the context of the evolutionary fulfillment of the human being in perfect enlightenment. For example, in the biology context, the way Buddha revised the pre-existing Indian theory of karma, which meant "fate" to some, and divine providence evoked by ritual action to others, it served as a proto-Darwinian theory of evolution. The relationship of human to animal life-forms was observed and argued, and what we consider to be ethically positive and negative was logically tied into the individual's actions of mind as well as body and speech. Most important, an individual was given a practical, biological reason to think and behave ethically, as every deed would have consequences in the rest of this life and throughout an open-ended individual, evolutionary future. The radical insight of Darwin about the genetic interconnectedness of all life-forms was clearly anticipated, with the additional element of the individual mind being involved, life after life, and not just the individual's genes as carried along in the development of the species. The denial of the individual person's responsibility for his or her own future lives seemed mandatory to Darwin in order for him to succeed to break away from the blind faith demand of dogmatic monotheism. This was not needed by the Buddha, and so he was able not only to reveal the interconnectedness of all life forms and to provide a plausible explanation for the differences of individuals and species, but he also was able to bind the individual ethically into positive actions through enlightened self-interest in pursuing a positive evolutionary trajectory within a relativistic cosmos that offered limitless opportunity and serious danger if wrongly pursued. This theory was acknowledged of course to be a hypothetical theory about relative reality, based on observation of the widely recognized ability of many individuals to remember their own previous lives, and not an absolute dogma demanding blind obedience. If anyone could have proved that there was no continuity of individual consciousness after the break up of brain and body, this would then have been accepted, and the theory revised.

But so far, no one, not then and not today, since none of even the most modern empirical scientists, has succeeded in proving that there is no continuity of consciousness once it has departed from the living brain. It is merely asserted dogmatically by scientific materialists. Observing a dead brain and not finding any energetic activity in it except for the processes of physical decomposition, does not prove that consciousness could not have continued outside of it. Smashing a computer with a sledge-hammer, when it could have sent all its files of information out across the internet, does not prove that the files have ceased to exist!

3. Some theorists maintain that science and religion occupy non-overlapping magisteria—i.e., that science and religion each have a legitimate magisterium, or domain of teaching authority, and these two domains do not overlap. Do you agree?

Not really, though this was a brave attempt by S. J. Gould to create grounds for a truce between the Dawkins types and the creationist types. I believe Western science has been too narrowed into dogmatic materialism since the time of Von Helmholtz and his "anti-vitalists oath," and Western religion has been too narrowed into supporting blind faith and irrationalism, as intimidated by the physical successes of science based technologies, even though the use of those technologies are destroying life on earth via industrial consumerism and industrial militarism.

4. What do you consider to be your own most important contribution(s) to theorizing about science and religion?

I don't think I've made any very important contribution at all, except insofar as I have made a little progress in elaborating what Arnold Toynbee meant when he said at Wellesley College in a speech in 1971 that the most important event of the 20th century would some day be considered to be the encounter of the West with Buddhism. Most who have heard of that surprising statement, as Toynbee was certainly not a Buddhist himself, have thought he might have been vaguely alluding to his expectation of a major missionary activity of Buddhism as an alternative religion, i.e. an inter-religious encounter. But if you connect the statement with his analysis of the Axial Age in Eurasia, time of Buddha, Isaiah, Confucius, Pythagoras etc., it is an inter-civilizational encounter, not a clash but an upgrade. Inter-religious impact is there of course, but also perhaps more importantly inter-scientific impact is enormous, as Western science focused on the mechanisms of material things, indeed has become dogmatically materialist, while Buddhist science always focused on what they called "inner," "psychological," "spiritual" science, not in a "mystical," but rather in an empirical, rational, manner. The meeting of the material and spiritual sciences are the core of this encounter, though it is still only getting under way.

5. What are the most important open questions, problems, or challenges confronting the relationship between science and religion, and what are the prospects for progress?

The most important thing is for the West to overcome its Euro-chauvinism and general intellectual arrogance, and realize that our conquest of the world over the last 500 years does not make us superior to the civilizations we conquered. Indeed, it might be a sign of our inferiority, our violent barbarity. When Gandhi responded to Churchill's question as to what he thought about Western civilization, the dhoti-clad Indian gentleman said, something to the effect of: "Western civilization? That would be a good idea!" Our sciences and religions enable us to conquer others, yes, but we have not conquered "nature," nor have we learned to understand or control our own selves, in terms of ethics, inner well-being, enjoying a meaningful sense of purpose in life, or even thinking there is any such thing. Thoreau said "the mass of men today lead lives of quiet desperation!" That is even more true today. I make these points to point out that we have a lot to learn from other civilizations.

How do we change our education systems to develop people's ethical sense, their altruism, their confidence in their own critical reason, their ability to learn from others, their mastery of several languages (Americans' language literacy is abysmal), their numeracy, and even their sense of hopefulness, contentment, and gentleness? Why should education only provide information and technical mastery. As the Dalai Lama said to the trustees and faculty of Columbia University when he received their doctorate of humane letters, "It is dangerous to educate only the clever brain without developing the good heart. How do you educate the good heart? That is the important question." Everyone nodded approvingly, but no one has any idea of how to do it, and I believe very little progress has been made in the decades since.

In short, in our universities, science is subordinated to commerce and considered the important activity on campus, and religion is considered to be vestigial, perhaps atavistic, something perhaps to be studied as phenomena by social scientists. How is there to be any progress in dialogue between them in such a climate? Fixing this fragmentation is a major challenge before us.

29

Michael Tooley

Michael Tooley is an American philosopher well known for his work in philosophy of science, philosophy of religion, metaphysics, and moral philosophy. He has taught at the University of Colorado Boulder since 1992 and was named College Professor of Distinction in 2006. Tooley is a Fellow of the Australian Academy of the Humanities and former President of the Australasian Association of Philosophy (1984) and the American Philosophical Association, Pacific Division (2010-11). His books include *Causation: A Realist Approach* (1987), *Time, Tense, and Causation* (1997), *Knowledge of God* (co-authored with Alvin Plantinga) (2008), and *Abortion: Three Perspectives* (co-authored with Alison Jaggar, Philip E. Devine, and Celia Wolf-Devine).

---

1. What initially drew you to theorizing about science and religion?

When I entered university, I was a Christian, and had no doubts at all about my religious beliefs. But then I had never really had any serious critical thoughts about any of my beliefs, religious or otherwise! In my first year at university, however, I had a conversation with a good friend, whose intelligence I very much respected, who suggested that I read Bertrand Russell's book *Marriage and Morals*. As I did so, I realized for the first time that one could ask whether one has good reasons for accepting the beliefs that one does. I then thought about my own religious beliefs, and I concluded that I didn't have good reasons for think-ing that they were true: they were simply beliefs that I had absorbed from my family, and from the church that I had joined as a young boy.

I then read other books by Russell in the areas of ethics and religion. One was his book *Religion and Science*, where Russell focused, on the one hand, on some historical conflicts between science and some traditional religious beliefs, including, for example, the conflict involving Copernicus' heliocentric view of the solar system, and the conflict involving Darwin's theory of evolution, and then, on the other hand, on the contrast between the ways in which religious beliefs are typically arrived at and held, as compared with the methods that lie at the very heart of science for arriving at and testing beliefs.

With regard to the former—the conflict between religious beliefs and scientific beliefs—one writer Russell referred to was Andrew White, whose massive book *A History of the Warfare of Science with Theology in Christendom*—it is almost 900 pages long—provided a very detailed overview of the extent to which many beliefs accepted at various times within just one religion—Christianity—had conflicted with scientific beliefs in an enormous number of areas, including not just creation versus evolution, the heliocentric theory of the solar system, the age of the Earth and the age of the universe, but also with scientific beliefs in such areas as geology, anthropology, ethnology, meteorology, medicine, anatomy, mental illness, and philology.

2. Do you think science and religion are compatible when it comes to understanding cosmology (the origin of the universe), biology (the origin of life and of the human species), ethics, and/or the human mind (minds, brains, souls, and free will)?

The formulation of this question is, I suggest, slightly unsatisfactory as it stands. First of all, while the term "science," naturally picks out a certain set of beliefs—namely, beliefs that are very highly confirmed within some science or other—the term "religion" does not do so: there are many religions in the world, with very different beliefs, and that vary greatly as regards compatibility with science. So no blanket answer to this question is possible.

Cosmology

Let us begin, however, with cosmology and the origin of the universe, focusing specifically on Christianity. Here everything may appear to turn upon one's view of the Bible. Suppose one is a fundamentalist, and believes that the Bible is completely free of error. Then one will hold that the account of creation found in *Genesis* 1:1 through *Genesis* 2:3 is correct. But according to that account, for example, the sky is a firmament in which the sun, the moon, and the stars are embedded. Moreover, those heavenly bodies did not arise via natural causal processes: rather, they were created directly by Yahweh. But both of these propositions are certainly incompatible with the relevant scientific theories.

Suppose, however, that one is not a fundamentalist. Then rather than holding that everything in the Bible has been revealed by God, one might hold that what has been revealed are truths that are spiritually important, including truths about the existence and nature of the God, about God's commandments, and about the nature and destiny of human beings. Since purely scientific facts are not important in that way, the deity could have remained silent on such matters, thereby giving the writers of the Bible the freedom to employ the sometimes scientifically faulty concepts of the day to express important spiritual truths. In particular, it might be argued that what is spiritually crucial as regards cosmology is simply what is affirmed in the great creeds of Christianity, such as the Apostles' Creed, and the Nicene Creed, to the effect that God the Father Almighty is the "creator of heaven and earth," the "creator of all things visible and invisible": whether the creation of our physical universe involved a single creative act, in which God caused the Big Bang, or, instead, involved several acts of special creation is of no religious importance, so there is no reason to assume that the Bible is accurate in its detailed account of exactly *how* God created the physical universe.

Can one conclude, then, that given this alternative view of the Bible, there need be no incompatibility between Christianity and science in the area of cosmology? That conclusion would, I think, be a hasty one. To see why, we need to consider a second, and more subtle point concerning the formulation of the second question. The starting point is with the claim that even the very best of current scientific theories involve propositions that have not been scientifically established. This may initially seem wildly implausible, but here are two examples. Consider Bishop Berkeley's view that there are no material objects, that reality consists only of God and finite immaterial minds, or the following, closely related view: "According to many schools of Hinduism, the world is an illusion, a play of the supreme consciousness of God. It is projection of things and forms that are temporarily phenomenal and sustain the illusion of oneness and permanence."[1] Theories in physics certainly appear to affirm the existence of a mind-independent material world, rather than the merely phenomenal world of Berkeley and many schools of Hinduism. But has physics established the existence of a mind-independent world?

A second example is this. Scientific theories involve claims about laws of nature, or at least about the existence of regularities, either exceptionless or probabilistic—such as that involved in the law of conservation of mass/energy. Has it been shown by physicists that it is at least likely that an exceptionless regularity concerning the conservation of mass/energy will hold in the future? To have done that, one must have solved the problem of justifying induction, and there is good reason to think that physicists have not done that. For one can show, for example, that if laws are viewed as mere regularities, without any more robust metaphysical backing, then two of the standard approaches to logical probability, one involving the assumption that all state descriptions are equally likely, and the other involving the assumption that all structure descriptions are equally likely, lead to the conclusion that one can never be justified in believing that there are any laws of nature, if laws are

---

[1]   Jayaram V., "The Definition and Concept of Maya in Hinduism," http://www.hinduwebsite.com/hinduism/essays/maya.asp, accessed January 10, 2014.

merely certain regularities, either exceptionless or probabilistic.

So theories in physics entail important propositions that have not been established scientifically, such as that there is a material world, rather than merely a phenomenal, mind-dependent one, and that certain regularities that have held in the past are likely to hold in the future.

But while these propositions have not yet been established, it would seem that they certainly could be. Thus, as regards the proposition that there is a mind-independent physical world, Frank Jackson, for example, has argued that it can be confirmed via hypothetico-deductive method, just as can be done for the molecular theory of gases,[2] while as regards the proposition that certain regularities that have held in the past will hold in the future, David Armstrong[3] and others have argued that if, rather than viewing laws of nature just as certain privileged regularities, one instead adopts what is known as a 'governing law' conception, according to which laws of nature are second-order relations between universals that, in the non-probabilistic case, entail corresponding regularities, then one can justify beliefs about the existence of laws via an inference to the best explanation.

My point here is not that these claims are clearly right, for while I do think that Jackson and Armstrong are right, I also think that **much** more needs to be done in both cases, involving rather heavy mathematics. My point is rather that if such lines of argument are right, then among the well confirmed propositions that physics will involve will be the propositions that there is a mind-independent world, rather than merely a phenomenal world, and that there are governing laws of nature that have not only held in the past but that also will hold in the future.

But what does this have to do with the relation between science and religion as regards cosmology? The answer lies in the evidential argument from evil, and the story begins with David Hume's *Dialogues Concerning Natural Religion*. There Hume considers different versions of the argument from evil. One is an incompatibility version of the argument from evil, according to which the existence of any evil is logically incompatible with the existence of an omnipotent, omniscient, and morally perfect being. But Hume then goes on to advance an evidential version of the argument from evil, the core of which is most clearly formulated in the following passage from near the end of Part XI:

---

[2] Frank Jackson, *Perception—A Representative Theory* (Cambridge, Cambridge University Press, 1977), Chapter 6.

[3] D. M. Armstrong, *What Is a Law of Nature?* (Cambridge: Cambridge University Press, 1983), pp. 104-6.

There may *four* hypotheses be framed concerning
the first causes of the universe: that they are endowed with perfect goodness; that they have perfect
malice; that they are opposite and have both goodness and malice; that they have neither goodness
nor malice. Mixed phenomena can never prove the
two former unmixed principles; and the uniformity
and steadiness of general laws seem to oppose the
third. The fourth, therefore, seems by far the most
probable.[4]

As I have argued elsewhere,[5] one has here a version of the evidential
argument from evil that appeals, in effect, to the idea of an inference to
the best explanation to decide between the competing hypotheses.

I favor a different approach to formulating an evidential argument
from evil, since I doubt that any principle of inference to the best explanation can possibly be a fundamental principle. The right approach,
I claim, involves bringing the idea of logical probability to bear upon
the issue, and I argue that when one does that, one can establish that if
there are *n* states of affairs in the world, where the *known* wrongmaking
properties of allowing each state of affairs to exist outweigh the *known*
wrongmaking properties of allowing that state of affairs to exist, then
the probability that *none* of those *n* states of affairs is such that the
totality of the wrongmaking properties, *both known and unknown*, of allowing that state of affairs to exist outweighs the totality of the wrongmaking properties, *both known and unknown*, of allowing that state of
affairs to exist is less than $\frac{1}{n}$. It then follows that, relative to the same
evidence, the probability that an omnipotent, omniscient and morally
perfect deity exists must also be less than $\frac{1}{n}$.[6]

Suppose, for the sake of discussion, that this probabilistic version of
the argument from evil is sound. Is it then a *scientific* truth that God
does not exist? It might be argued that it is not, on the ground that

---

[4] David Hume, *Dialogues Concerning Natural Religion*, ed. Richard H. Popkin
(Indianapolis, Indiana: Hackett Publishing Company, 1980), p. 75.

[5] Michael Tooley, "Hume and the Problem of Evil," in Jeffrey J. Jordan (ed.)
Philosophy of Religion: The Key Thinkers (London: The Continuum International
Publishing Group, 2011), pp. 159-86.

[6] Slightly different proofs of this result are set out in Michael Tooley, "Does God
Exist?" in Alvin Plantinga and Michael Tooley, *Knowledge of God* (Oxford: Oxford University Press, 2008), pp. 70-150, and in Michael Tooley, "Inductive Logic
and the Probability that God Exists: Farewell to Skeptical Theism," in *Probability
in the Philosophy of Religion*, edited by Jake Chandler and Victoria S. Harrison,
Oxford, Oxford University Press, 2012, 144-64.

scientific reasoning, rather than appealing to principles of logical probability, makes use of methods such as hypothetico-deductive reasoning and inference to the best explanation. I would argue, however, that neither of those methods is a satisfactory candidate for a basic principle of inductive logic, and that, to the extent that those methods are sound, this must be established by appealing to principles of logical probability.

It can also be objected, however, that it cannot be a scientific truth that God does not exist, since science does not make any claims about moral properties, and God is by definition morally perfect. This objection, in contrast to the first, has real force, but it can be finessed as follows. In judging that evils are present in the world, one must appeal to some set of moral standards. So let M be the moral standards used. One can then parallel the probabilistic argument from evil in question to arrive at the following conclusion:

> P: The probability that there is an omnipotent and omniscient deity that is morally *perfect given moral standards* M is less than $\frac{1}{n}$.

This conclusion, however, in contrast to the earlier one, does not assume that moral standards M are correct, or even that there are any correct moral standards. The claim that science does not make any claims about moral properties is therefore no objection to the claim that *P* is, if true, a scientific proposition.

To sum up, initially it might seem that a Christian can avoid a collision course with scientific beliefs about the origin of the universe by rejecting a fundamentalist view of the Bible, and holding that what is religiously crucial is that the Christian deity, an all-powerful being, created our spatiotemporal world, and that it is a matter of no importance exactly how that was done. But the problem is, first of all, that any Christian will surely claim that the creator of our universe, rather than being morally indifferent, or evil, is good, indeed perfectly so; and, secondly, if M is the set of moral principles that a given Christian accepts, and if there is even one state of affairs, S, in our world that, judged by the principles in M, is such that the known wrongmaking properties of an action of allowing S to exist outweighs the known rightmaking principles of doing so, it will then follow that the probability that there is an omnipotent and omniscient being that is also morally perfect, given the moral standards in M, is at the very least less than $\frac{1}{2}$.

In short, if a certain version of the argument from evil is sound, there is a proposition that, first of all, although not part of any current science, can be established by principles that are needed to justify scientific methods of confirming propositions, and secondly, is incompatible with a core Christian claim—namely, that our universe was created by a being who, judged by certain relevant moral standards, is perfectly good.

Biology

What about the origin of life, and the origin of the human species? As regards the former, there is no incompatibility at present, since there is no well-confirmed scientific theory of how life arose from inanimate things.

What about the origin of the human species? If one holds a fundamentalist view of the Bible, one believes that human beings were specially created by Yahweh, as described in Genesis 2:7 and Genesis 2:21-22. This view is on a collision course with the view that humans are descended from non-human primates, for which there is massive DNA evidence, as is well described in Daniel J. Fairbank's book *The Relics of Eden* (Amherst, New York: Prometheus Books, 2007), and which includes such things as: (1) the relation between two of the chromosomes found in non-human primates, and chromosome number two in the case of humans—a relation that can only be plausibly explained by the hypothesis that the latter resulted from the former via fusion; (2) facts involving transposable elements, which provide evidence for the hypothesis that human DNA arose out of the DNA of earlier, non-human primates; and (3) the relations between pseudogenes (bogus genes) found in humans on the one hand, and non-human primates on the other.

The view that humans arose via a special act of creation is, then, untenable. This leaves one with two alternatives: either humans arose from non-human primates by a completely naturalistic process of evolution, or else humans originated via divine intervention in what would otherwise be a purely naturalistic process. The question is then whether the purely naturalistic hypothesis is well confirmed, or whether the alternative hypothesis of theistically guided evolution is not equally plausible as things presently stand.

There are two main objections to the theistic alternative. First, there is once again the evidential argument from evil, both the general argument, and a version that focuses specifically on 'design faults' present in human beings, including such things as the human spine, sinuses, wisdom teeth, the size of the birth canal, different susceptibility of men and women to lung cancer, hormone levels that decline with age, and many others that I have listed elsewhere.[7]

---

[7] Michael Tooley, *Knowledge of God*, pp. 110-13.

Secondly, the plausibility of any view that involves supernatural intervention in the natural world is affected by the extent to which there are instances elsewhere where it is plausible to hold that supernatural intervention has occurred. Is there good reason to believe, then, that miracles have occurred?

This is an enormous issue, but what I would argue is that there are a number of studies that make it plausible that miracles do not occur, including Andrew D. White, "The Growth of Healing Legends," in *A History of the Warfare of Science with Theology within Christendom* (Buffalo, New York: Prometheus Books, 1993), D. J. West, *Eleven Lourdes Miracles* (London: Gerald Duckworth and Company, 1957), Louis Rose, *Faith Healing* (London: Victor Gollancz, 1968), William A. Nolen, *Healing: A Doctor in Search of a Miracle* (New York: Random House Inc., 1974), together with, for example, the famous STEP Project (Study of the Therapeutic Effects of Intercessory Prayer).

The Human Mind

Many religions involve, as a central element, the belief that humans have an immaterial mind or soul that is, among other things, the ground of personal identity. This belief is crucial, for example, to the belief in reincarnation that one finds in Hinduism and Buddhism, but is also central to the account of human nature advanced by Roman Catholicism, and accepted by many Protestants as well.

Within contemporary psychology, by contrast, there is massive acceptance of the view that the human mind is identical with the human brain, and thus is a purely material entity. So we have religious views that are incompatible with a currently dominant scientific view of the human mind.

But are there good reasons to accept that psychological view? Elsewhere I have argued that there are a number of facts that provide excellent evidence for the view that the categorical bases of human psychological capacities lie not in some immaterial entity, but in complex neural structures and circuitry.[8] These include, among many others: (1) the fact that minor blows to the head can cause unconsciousness; (2) the correlations between damage to different parts of the brain and the impairment of specific psychological capacities; (3) the effects of diseases such as Alzheimer's; (4) the effects of aging on mental capacities; (5) the effects of psychotropic drugs on states such as anxiety and depression; (6) the inheritance of personality traits, including intelligence. I

---

[8] Michael Tooley, "4. The Appeal to an Immaterial, Rational Mind," in Michael Tooley, Celia Wolf-Devine, Philip E. Divine, an Alison M. Jaggar, *Abortion: Three Perspectives* (New York: Oxford University Press, 2009), pp. 15-19.

believe, then, that the hypothesis that the mind is not an immaterial entity is one that is very well confirmed.

Ethics

I have left this area to the last, since while some present day philosophers believe that ethical truths can be scientifically established, by far the dominant view is that this is not the case, and the reason offered is that while propositions established by science are merely contingently true, the most basic principles in ethics are, if true, necessarily true.

Regardless of one's view on this issue, however, one can ask whether present day ethical views are compatible with religious views. Once again, let us focus specifically on Christianity. There it seems to me that there are serious incompatibilities. Some of these concern central tenets of Christianity, such as the doctrines of original sin, of an atoning sacrifice, and of hell. Thus very few major philosophers working in the areas of ethics would accept any of the following claims: (1) It is morally permissible for all humans to be made worse off by something that was done by the first human; (2) In the absence of an atoning sacrifice by a person who was both God and man, it would be morally impossible for God to forgive humans their sins; (3) It is morally permissible for some people—or even the majority of the human race—to undergo eternal torment.

Other incompatibilities include various moral principles. Thus Jesus taught that premarital sexual intercourse was something that defiled a person, and that divorce was permissible at most in the case of adultery, while Paul held that homosexual sex was not only morally wrong, but seriously so. Very few philosophers working in the area of ethics, and who are not religious, accept these claims.

There are, in short, serious incompatibilities between central Christian tenets and moral teachings and the views of the vast majority of contemporary philosophers working in the area of ethics.

3. Some theorists maintain that science and religion occupy non-overlapping magisteria—i.e., that science and religion each have a legitimate magisterium, or domain of teaching authority, and these two domains do not overlap. Do you agree?

The idea that science and religion occupy non-overlapping magisteria was made famous by Stephen Jay Gould, who defended it at length in his book *Rocks of Ages: Science and Religion in the Fullness of Life* (New York: Ballantine Books, 2002). But don't most religions make claims about the nature and origin of the universe, the origin of life and the human species, the nature of man, and human destiny—and also, in the case of some religions, crucial historical claims about, for example, Moses, or Jesus, or Mohammad—and aren't those claims such as fall within the scope of scientific investigation, broadly understood to include history? So how can Gould's non-overlapping magisteria claim be true?

Gould's strategy in defending this claim involved the following contention: "The magisterium of religion extends over questions of ultimate meaning and moral value" (2002, 6). But isn't this to define "religion" in a way that reduces religion to ethics, and that leaves out most of the elements that religious believers view as absolutely crucial to the religions that they accept? It seems to me, then, that Gould's idea is quite unsound.

Notice in addition, however, that even if one adopted an extraordinarily liberal view of Christianity that included only the ethical elements of traditional Christianity, one would still not have non-overlapping magisteria, and for two reasons. First, ethics is part of philosophy, and has been since philosophy began with Socrates, so there would be overlap, and very likely conflict there. Secondly, what grounds could there be for assigning any authority to 'religious' views, thus understood? For what method, distinct from those of philosophy, could then serve to justify ethical claims?

4. What do you consider to be your own most important contribution(s) to theorizing about science and religion?

First, as discussed above, I believe that my version of the evidential argument from evil, based as it is on principles of logical probability that I think are needed for the justification of scientific methods—such as inference to the best explanation—is a *scientific* argument against the existence of God.

Secondly, in my article "Plantinga's New Argument against Materialism,"[9] I argued that philosophers who attempt to show that the mind is an immaterial entity generally fail to address the strongest objection against that view, an objection that I set out in a fairly detailed way.

Finally, in an article entitled "Naturalism, Science, and Religion,"[10] I discuss a number of arguments that have been offered in support of supernaturalism, both traditional and more recent ones, and I argue that none of those arguments provides a good reason for thinking that supernaturalism is true.

---

[9] Michael Tooley, "Plantinga's New Argument against Materialism," *Philosophia Christi*, 12/1, 2012: 27-45.

[10] Michael Tooley, "Naturalism, Science, and Religion," in Bruce L. Gordon amd William A. Dembski, *The Nature of Nature* (Wilmington, Delaware: ISI Books, 2011), pp. 880-900.

5. What are the most important open questions, problems, or challenges confronting the relationship between science and religion, and what are the prospects for progress?

As a person who thinks that all religions involve core beliefs that there are excellent reasons, either scientific or philosophical, for thinking are false, the crucial challenge that I see is not an intellectual one, but an educational one, and one that is especially great in the United States, where a very large number of people still believe that humans were specially created by a deity in the last few thousand years. We have a society where, if one does not go on to university, it is very unlikely that one will be exposed even to evidence bearing upon the age of the Earth, let alone the DNA evidence for human evolution, or arguments for and against the existence of God, or philosophical criticisms of religious answers to controversial ethical questions concerning sexuality, marriage, abortion, euthanasia, and many other topics. Until that is changed, whatever progress those of us who are scholars make in getting into sharper focus questions about the relation between science and religion is unlikely to enable ordinary people to arrive at views on these important matters that will withstand critical scrutiny.

30

Charles Townes

Charles Townes is an American Nobel Prize-winning physicist and one of the leading figures in twentieth-century physics. He is the inventor of the maser, co-inventor of the laser, and a pioneer in microwave spectroscopy for molecular and nuclear physics and in the use of radio and infrared spectroscopic techniques for astronomy. He was one of the first academic scientists to accept a full-time position advising the Executive Branch during the Cold War, and was founder of the "Jasons," an influential group of scientists independently advising the government. In 1964, he shared the Nobel Prize in Physics with Nikolay Basov and Alexander Prokhorov for their ground breaking work on quantum electronics. And in 2005, he was awarded the Templeton Prize for "Progress Toward Research or Discoveries about Spiritual Realities." A member of the United Church of Christ, Townes sees science and religion as parallel and more similar than most people think. He is the author of *How the Laser Happened: Adventures of a Scientist* (2000) and *Making Waves* (1995).

---

1. What initially drew you to theorizing about science and religion?[1]

Well, both are very important in our lives and I want to understand them as well as I can.

Even as a youngster, you're usually taught that there's some purpose you'll try to do, how you are going to live. But that's a very localized thing, about what you want with your life. The broader question is, "What are humans all about in general, and what is this universe all about?" That comes as one tries to understand what is this beautiful world that we're in, that's so special: "Why has it come out this way? What is free will and why do we have it? What is a being? What is consciousness?" We can't even define consciousness. As one thinks about these broader problems, then one becomes more and more challenged by the question of what is the aim and purpose and meaning of this universe and of our lives.

Those aren't easy questions to answer, of course, but they're important and they're what religion is all about. I maintain that science is closely related to that, because science tries to understand how the uni-

---

[1] This interview is derived from two sources: a phone interview conducted by the editor of this volume and a previously published interview conducted by Bonnie Azab Powell for the *UC Berkeley NewsCenter* (June 17, 2005). Previously published material is reprinted here with the permission of the *UC Berkeley NewsCenter* and Charles Townes. The final interview, as it appears here, has been approved by Charles Townes.

verse is constructed and why it does what it does, including human life. If one understands the structure of the universe, maybe the purpose of man becomes a little clearer. I think maybe the best answer to that is that somehow, we humans were created somewhat in the likeness of God. We have free will. We have independence, we can do and create things, and that's amazing. And as we learn more and more—why, we become even more that way. What kind of a life will we build? That's what the universe is open about. The purpose of the universe, I think, is to see this develop and to allow humans the freedom to do the things that hopefully will work out well for them and for the rest of the world.

2. Do you think science and religion are compatible when it comes to understanding cosmology (the origin of the universe), biology (the origin of life and of the human species), ethics, and/or the human mind (minds, brains, souls, and free will)?

Science and religion have had a long interaction: some of it has been good and some of it hasn't. As Western science grew, Newtonian mechanics had scientists thinking that everything is predictable, meaning there's no room for God—so-called determinism. Religious people didn't want to agree with that. Then Darwin came along, and they *really* didn't want to agree with what he was saying, because it seemed to negate the idea of a creator. So there was a real clash for a while between science and religions.

But science has been digging deeper and deeper, and as it has done so, particularly in the basic sciences like physics and astronomy, we have begun to understand more. We have found that the world is not deterministic: quantum mechanics has revolutionized physics by showing that things are not completely predictable. That doesn't mean that we've found just where God comes in, but we know now that things are not as predictable as we thought and that there are things we don't understand. For example, we don't know what some 95 percent of the matter in the universe is: we can't see it—it's neither atom nor molecule, apparently. We think we can prove it's there, we see its effect on gravity, but we don't know what and where it is, other than broadly scattered around the universe. And that's very strange.

So as science encounters mysteries, it is starting to recognize its limitations and become somewhat more open. There are still scientists who differ strongly with religion and vice versa. But I think people are being more open-minded about recognizing the limitations in our frame of understanding.

Now, with regard to the origin of life and so on, science and religion both fit those very well. Both can apply to that.

I do believe in both a creation and a continuous effect on this universe and our lives, that God has a continuing influence—certainly his laws guide how the universe was built. But the Bible's description of creation occurring over a week's time is just an analogy, as I see it. The Jews couldn't know very much at that time about the lifetime of the universe or how old it was. They were visualizing it as best they could and I think they did remarkably well, but it's just an analogy.

As for the debate over Intelligent Design in the United States, I think it's very unfortunate that this kind of discussion has come up. People are misusing the term intelligent design to think that everything is frozen by that one act of creation and that there's no evolution, no changes. It's totally illogical in my view. Intelligent design, as one sees it from a scientific point of view, seems to be quite real. This is a very special universe: it's remarkable that it came out just this way. If the laws of physics weren't just the way they are, we couldn't be here at all. The sun couldn't be there, the laws of gravity and nuclear laws and magnetic theory, quantum mechanics, and so on have to be just the way they are for us to be here.

Some scientists argue that "well, there's an enormous number of universes and each one is a little different. This one just happened to turn out right." Well, that's a postulate, and it's a pretty fantastic postulate— it assumes there really are an enormous number of universes and that the laws could be different for each of them. The other possibility is that ours was planned, and that's why it has come out so specially. Now, that design could include evolution perfectly well. It's very clear that there is evolution, and it's important. Evolution is here, and intelligent design is here, and they're both consistent.

People who want to exclude evolution on the basis of intelligent design, I guess they're saying, "Everything is made at once and then nothing can change." But there's no reason the universe can't allow for changes and plan for them, too. People who are anti-evolution are working very hard for some excuse to be against it. I think that whole argument is a stupid one. Maybe that's a bad word to use in public, but it's just a shame that the argument is coming up that way, because it's very misleading.

3. Some theorists maintain that science and religion occupy non-overlapping magisteria—i.e., that science and religion each have a legitimate magisterium, or domain of teaching authority, and these two domains do not overlap. Do you agree?

No.

4. What do you consider to be your own most important contribution(s) to theorizing about science and religion?

I'm not sure I made a lot of contributions. I think about the relationship between science and religion, talk about it. But have I made contributions different from the ones others have made? I don't know that I have. If anything, my contribution has been the recognition of their parallelism and consistence.

5. What are the most important open questions, problems, or challenges confronting the relationship between science and religion, and what are the prospects for progress?

Well, one important question is: How did it all begin? That is a very important question. Science cannot answer that at all. Religion does, of course. How did it all begin?

Another question is: Why are the laws of science the way they are? Why do they allow us to be here? How did it turn out that way? They had to be special for us to be here. So there are a lot of interesting questions.

Any final thoughts?

Science attempts to understand how things work. Religion attempts to understand why they are the way they are. They kind of fit together, but they are not the same thing.

31

Peter van Inwagen

Peter van Inwagen is an American philosopher and one of the leading figures in contemporary metaphysics and philosophical theology. He is currently the John Cardinal O'Hara Professor of Philosophy at the University of Notre Dame. Van Inwagen was elected to the American Academy of Arts and Sciences in 2005, was President of the Central Division of the American Philosophical Association in 2008/09, and was President of the Society for Christian Philosophers from 2010 to 2013. His books include *An Essay on Free Will* (1983), *Material Beings* (1990), *Metaphysics* (1993), *God, Knowledge and Mystery* (1995), *The Possibility of Resurrection and Other Essays in Christian Apologetics* (1997), *Ontology, Identity, and Modality* (2001), *The Problem of Evil* (2006), and *Existence: Essays in Ontology* (forthcoming). In 2011, van Inwagen was awarded the degree of Doctor of Divinity *honoris causa* from the University of St. Andrews.

1. What initially drew you to theorizing about science and religion?

First, I have to say that I don't think it's very accurate to say that I have ever done very much of something called "theorizing about science and religion." What I mean by this will become clear in the course of my successive answers to your questions—at least I hope so. One way to answer Question 1 without raising questions about the meaning of the phrase 'theorizing about science and religion' would be to say that (1) I am a Christian, and that Christianity is often under attack; that (2) I have responded to some of these attacks for more or less the same reason that people write letters to the *New York Times* responding to the arguments of some editorial or op-Ed piece; and that (3) some of these attacks have—at least this was the intention of the people making the attacks—been based on science.

2. Do you think science and religion are compatible when it comes to understanding cosmology (the origin of the universe), biology (the origin of life and of the human species), ethics, and/or the human mind (minds, brains, souls, and free will)?

In my view, it's not useful to ask whether "religion" is compatible with anything. What falls under the umbrella-term 'religion' is simply too various. One might as well ask whether "politics" is compatible with understanding cosmology and biology as ask whether "religion" is compatible with understanding cosmology and biology. Obviously, Stalin's politics (with its endorsement of Lysenkoism) and Hitler's politics (which required German scientists solemnly to testify that "race" in the Nazi sense—"the Aryan race," "the Slavic race"—was an objectively real biological category) are not compatible with scientific biology. And, just as obviously, membership in many political parties or groups or schools of thought is compatible with an understanding of biology. (As to politics and cosmology, consider the Nazi attempt to suppress the "Jewish physics" of Albert Einstein—that is, the general theory of relativity, an essential tool of modern cosmology—or the Soviet opposition to "big bang" cosmology, owing to the fact that a past-eternal material universe is an axiom of dialectical materialism.)

Questions like the following make more sense than 'is religion compatible with science' (or '... with cosmology,' '... with biology' and so on). Is the Christian doctrine of creation compatible with the steady-state cosmology? Is the Hindu doctrine of the transmigration of souls compatible with what we know about the relation between neurobiology and the mind? (The former question displays the fact that it is not only the term 'religion' that applies to too diverse a class of things for it to be useful to ask whether its referent is compatible with something. 'Science' and 'cosmology' cover a lot of very diverse things as well. One scientific cosmology may be compatible with a certain theological position and another incompatible with that position. And this is not a mere possibility. I would say that the more-or-less uncontroversial parts of present-day cosmology are compatible with the Christian doctrine of Creation, whereas Hoyle's steady-state cosmology was not. But even that rather elaborately qualified assertion cries out for further qualification, since it raises the question of what Christian doctrines are to be included in "the Christian doctrine of Creation." According to Thomas Aquinas, at least, and I see no reason to dispute his arguments for this thesis, God had the power to create a cosmos with an eternal past. My statement that the steady-state cosmology is incompatible with the Christian doctrine of Creation assumes that that doctrine includes the proposition that the cosmos that God has *in fact* created began to exist at a particular moment.)

In the remainder of this answer, therefore, I will address only questions about the compatibility of the doctrines of my own religion, Christianity, with the currently accepted findings of cosmology and biology and various other sciences. And perhaps I should say, my own *version* of Christianity—"mainstream Christianity" I'll call it (perhaps somewhat tendentiously). Some versions of Christianity—I am thinking specifically of "six-days-of-creation biblical literalism"—are obviously incompatible with the findings of cosmology and biology. But, like St. Augustine and St. Thomas Aquinas, I am not a six-day literalist. I take six-day literalism—commonly called fundamentalism—to be a minor and theologically uninteresting offshoot of "mainstream" Christianity. I can't resist inserting at this point an anecdote based on my own experience. A former "fundamentalist" who had become an aggressive atheist once accused me of being a "wishy-washy theological liberal" because I believed that the earth was billions of years old and that any two terrestrial organisms had a common ancestor. I asked him whether he would say that St Augustine was a wishy-washy theological liberal. (I had just described Augustine's account—in *De Genesi ad litteram*—of Creation and his reconciliation in that book of that account with the creation story set out in first three books of Genesis.) "Oh, sure," he replied.

In about the year 1200, one Peter of Cornwall, Prior of Holy Trinity, Aldgate, wrote the following words:

> There are many people who do not believe that God
> exists, nor do they think that a human soul lives on
> after the death of the body. They consider that the
> universe has always been as it is now and is ruled
> by chance rather than by Providence.

(As far as I know, Peter's ms. is unpublished. The above passage was quoted in Robert Bartlett in *England under the Norman and Angevin Kings, 1075-1225* (Oxford: at the Clarendon Press, 2000). I presume the translation is the author's. I have copied it from a review of the book by John Gillingham in *The Times Literary Supplement* (5 May 2000, p. 26).)

Here is a second quotation from Bartlett's book—the words are the author's—by the reviewer (no page citations are given in the review):

> [S]imple materialism and disbelief in the afterlife
> were probably widespread, although they leave
> little trace in sources written by clerics and monks.

Now obviously science has learned a great deal since the year 1200. And what science has learned has shown that much of what educated people believed in that remote year was wrong: the earth is not at the center of the universe, the sun does not revolve around the earth, the fixed stars are not all at the same distance from us, the universe is not five thousand years old but *millions* of thousands of years old ... That is to say, astronomy and cosmology have shown that many—most—of the astronomical and cosmological beliefs of the medievals were false.

(Not to mention their beliefs about zoology and geography and history.) But has science shown that any of the religious beliefs of medieval Christians were false—or even doubtful? Or put the question this way. Consider those "many people" whose deplorable beliefs Peter of Cornwall describes in the above passage. Has science discovered anything since their day that gives aid and comfort to them and their world-view? If one of those long-dead unbelievers could be returned to life today and given a modern scientific education, would he find anything among the scientific discoveries of the last eight hundred years that would lead him to crow triumphantly, "Aha! I was right all along! I knew it!"

Not in my view. In my view, the discoveries of science are irrelevant to the question whether God exists (and are therefore irrelevant to the question whether the universe is ruled by chance or Providence). And, if modern cosmology teaches us anything, it teaches us that the universe has *not* always been as it is now. It might be argued, and I think very plausibly, that modern neurobiology, if its findings do logically not imply the non-existence of a separable soul—an immaterial entity that is the seat of our thought and experience—, has shown that belief in such a separable soul faces grave difficulties. (And what part or aspect of us could be supposed to be a soul that "lives on after the death of the body" if it were not an immaterial entity that was the seat of our thought and experience?) I do not, however, believe that belief in a separable soul is essential to Christianity. I will return to this point in my answer to question 3.

In my view, modern science has discovered nothing that refutes, or even casts any doubt on, any essential Christian doctrine. I would say, in fact, that only two of the many things science has learned in the last eight centuries are so much as *relevant* to any theological question.

One of them, I have already mentioned: we know today that the universe has *not* always been as it is now. Proponents of naturalism and materialism have long made their peace with this discovery, but they certainly resisted it very vigorously at first. (Consider Einstein's attempt to fiddle with the field equation of his new theory of general relativity to allow a "static" universe; consider the steady-state theory; consider the words of the Nobel Laureate physical chemist Walther Nernst: "To deny the infinite duration of time would be to betray the very foundations of science.")

Secondly, we know today that there were sentient animals long before there were rational animals. This greatly complicates the theological problem of "animal suffering." If it were possible to believe that human beings and sentient but non-human animals came into existence at more or less the same time—within the same six-day period, for example—, it would be possible to ascribe the sufferings of animals (the sufferings due to predation and parasitism and other factors not directly caused by the actions of human beings), to a corruption of nature by the abuse of human free will.

I know of no other scientific discoveries that are relevant to theology. (Some might want to add the "fine-tuning for life" of the parameters in the laws of physics and the boundary conditions of the cosmos. I would not include "fine-tuning" in my list, but I cannot explain why within the scope provided by these questions.)

3. Some theorists maintain that science and religion occupy non-overlapping magisteria—i.e., that science and religion each have a legitimate magisterium, or domain of teaching authority, and these two domains do not overlap. Do you agree?

The "non-overlapping magisteria" thesis is not as clear as it might be, and I'm not sure whether everyone who claims to subscribe to this thesis means the same thing by the words 'non-overlapping magisteria.' It seems to me that *most* people who claim to subscribe to this thesis take it to mean that it is the business of "religion" to answer questions about morals or about how people should live their lives or about what "meaning" is to be found in the human condition—and nothing else. And (they maintain) it is the business of science to settle questions about matters like the size and age of the universe and the history of life on the earth—and that science (though it certainly provides information that must be taken account of when one is seeking the answers to questions about morals, etc.) has overstepped its bounds if it pronounces on morals or how to live one's life or the meaning of our existence. (Those who say this do not, of course, deny that individual *scientists* may put forward theses on these matters. Anyone, the scientist included, has the right to do that, but scientists who exercise that right should state explicitly that they are speaking as private individuals—as citizens exercising their right of free speech—and should not claim to be speaking in the name of science.)

Some of this is no doubt right, but I don't think that, taken as a whole, the "non-overlapping magisteria" view is acceptable. My reason for saying this is not profound; it consists simply in a recognition of the fact that most religions—my own included—incorporate doctrines that pertain to matters other than morals and the way we should live our lives and the meaning of those lives. Here is one example that I have already touched on: if the steady-state cosmology had turned out to be strongly supported by the cosmological evidence, that support would have raised an important obstacle to belief in the Christian doctrine of creation. (I do not claim that my religion is in principle immune to scientific refutation; I claim only that if this is going to happen, it has not happened yet.) It seems to me that modern neurobiology has raised grave difficulties for the Hindu doctrine of the transmigration of souls. (But whether this is really so is a question I will leave to Hindu philosophers and scientists and theologians.) I will note, incidentally, that in my view, my own religion, Christianity, is not—as I believe Hinduism is—*essentially* committed to any sort of mind-body dualism. I grant that many, perhaps most, Christians *are* mind-body dualists. I think that the findings of neurobiology show that these Christians should stop being mind-body dualists. But whether I am right about that or not, my position is that they could cease being mind-body dualists without ceasing to be fully orthodox Christians.

4. What do you consider to be your own most important contribution(s) to theorizing about science and religion?

I think that there are three issues I've written on in which what I have had to say might be of some value. What I have said about all three of them resists compression—that is, it would be impossible to summarize what I have said about them in the scope allowed by this format in a way that would not be misleading. I will therefore have to refer the reader who would like to know more about my views on these matters to some lectures and essays in which those views are carefully stated.

The first has to do with what W. K. Clifford has called "the ethics of belief." Clifford summarized his ethics of belief in the following famous words:

> It is wrong always, everywhere, and for anyone, to
> believe anything upon insufficient evidence.

This ethico-doxastic principle—let us call it "Clifford's Principle"—is commonly held to be one of the fruits of modern science, which has taught a favored few among us the following lesson (the description of the lesson is by Thomas Henry Huxley):

> [Science] warns me to be careful how I adopt a
> view which jumps with my preconceptions, and to
> require stronger evidence for such belief than for
> one to which I was previously hostile. My business
> is to teach my aspirations to conform themselves to
> fact, not to try and make facts harmonise with my
> aspirations.

I have tried to show, first, that *no one*—neither the "man on the Clapham omnibus," nor the scientist, nor the philosopher, nor the Pope and the Archbishop of Canterbury—consistently applies Clifford's Principle in forming his or her beliefs. (Nor, I contend, have W. K. Clifford and T. H. Huxley themselves consistently applied this principle to their own beliefs.) And, I have tried to show in that essay, secondly, that the beliefs of the Bishop of Rome and his Anglican counterpart concerning supernatural matters fare no worse when judged according the standard Clifford has set than the beliefs of physical anthropologists concerning the evolutionary history of our species or the beliefs of astronomers about the source of "gamma-ray bursts." Clifford's ethical injunction may be valid, but—like another famous moral injunction, which has an excellent claim to validity: "Sell all that you have and give to the poor"—it is simply not a principle that anyone actually follows.

I have discussed Clifford's Principle in several essays. Perhaps the most accessible—the least technical—is "'It Is Wrong, Everywhere, Always, and for Anyone, to Believe Anything upon Insufficient Evidence'," in Jeff Jordan and Daniel Howard-Snyder (eds.), *Faith, Freedom, and Rationality*, (Rowman & Littlefield, 1996). (I note with embarrassment that in giving the essay that title, I misquoted Clifford's statement of his principle.)

I mentioned three "issues." The second is the relation between the Darwinian theory of evolution and belief in God (on the one hand) and the account of Creation in the first three chapters of Genesis (on the other). One of the places in which I have discussed evolution and God and Scripture is in a series of three lectures, "A Kind of Darwinism," "Darwinism and Design," and "Science and Scripture," which are printed in volume 2 of Melville Y. Stewart (ed.), *Science and Religion in Dialogue* (Wiley-Blackwell, 2010).

The last of the three issues is the increasingly popular project of providing "deflationary" evolutionary explanations of belief in the supernatural. Here I would direct readers to my "Explaining Belief in the Supernatural: Some Thoughts on Paul Bloom's 'Religious Belief as an Evolutionary Accident'" in Jeffrey Schloss and Michael Murray (eds.), *The Believing Primate: Scientific, Philosophical, and Theological Reflections on the Origin of Religion* (Oxford: Oxford University Press, 2010).

My essay on "Clifford's Principle," the three lectures on God and evolution, and the response to Professor Bloom's paper are all available online, at a site (maintained by my former student Andrew Bailey) where most of my work (other than my books) can be found in easily accessible form: http://andrewmbailey.com/pvi/

5. What are the most important open questions, problems, or challenges confronting the relationship between science and religion, and what are the prospects for progress?

In my view, there are two areas in which there is room for more, and better-informed, scholarly discussion. (As regards some topics, particularly "cosmology and fine-tuning" and "Darwinism and design in nature," I doubt whether—absent some new access of scientific knowledge—any work on these topics that is published in the future is going to contain any thoughts that aren't trivial variants on thoughts that have already appeared in print many times.)

I have mentioned one of these "areas" in my answer to the previous question: evolutionary explanations of belief in the supernatural.

The other is the question of the role the Church has played in the development of science. There used to be a story ("the Andrew White story," so to call it) about the relation between science and the Church, a story according to which the "adolescent" science of the classical world—geometry, astronomy, statics—was about to develop into "adult" science, science in in the fully modern sense; the science of the Greeks and Romans, however, was destroyed by the rise of Christianity (as part and parcel of Christianity's larger task of destroying the civilization of the Roman empire and issuing in the Dark Ages). The Church had its way for many centuries, and then—somehow—a light pierced the darkness. Somehow, in priest-ridden Europe, of all places, learning was reborn—and, along with many other branches of learning, physical science was reborn. (Why *there*? And, *if* there, why not also in the Arab world or in China or in India? Excellent questions both.) And, having achieved rebirth, science grew to the adulthood that the Church had been able to deny it in the ancient world. The Church, of course, tried to destroy the new science as she had destroyed the old (Galileo and all that), but—again, *somehow*—failed, and the Church has been losing ground to science ever since.

I don't think very many people take the Andrew White story seriously any more—not at any rate in the primitive form in which I've told it. But a lot of people seem still to believe latter-day more sophisticated (or vaguer) versions of it.

There is a rival story. According to the rival story, Western Latin Christianity was an essential causal element in the development of modern science—a thesis that, if true, explains why modern science (the science whose history includes intellectual achievements of an order unknown to the ancients, achievements such as Newton's deduction of Kepler's laws of planetary motion from his laws of motion and his law of universal gravitation) arose *only* in "priest-ridden Europe."

While I have in my work made a few short, scattered remarks about the role of the Church in the development of modern science, the history of science and the history of ideas are areas in which I am a rank amateur, and those remarks amounted only to a superficial presentation of ideas I had borrowed from real historians. I can do no more than refer the interested reader to the *locus classicus* of the line of thought I have briefly sketched, Stanley L. Jaki's Gifford Lectures (1974-76), published as *The Road of Science and the Ways to God* (Chicago: the University of Chicago Press, 1978). As far as I know, no one has continued Fr Jaki's work in this area. I very much hope that someone will.

32

Keith Ward

Keith Ward is a British philosopher, theologian, pastor and scholar, ordained priest of the Church of England, and Fellow of the British Academy. He specializes in comparative theology and the relationship between science and religion. He has held academic positions as Dean of Trinity Hall, Cambridge, F.D. Maurice Professor of Moral and Social Theology at the University of London, Professor of History and Philosophy of Religion at King's College London, Regius Professor of Divinity at Oxford, and visiting professor at Claremont Graduate University. He is the author of numerous books including *Ethics and Christianity* (1972), *In Defence of the Soul* (1992), *Comparative Theology* (5 volumes), *The Case for Religion* (2004), *Is Religion Dangerous?* (2006), *The Big Questions in Science and Religion* (2008), *Why There Almost Certainly Is a God* (2008), *The God Conclusion* (2009), *More than Matter: What Humans Really Are* (2010), *The Philosopher and the Gospels* (2011), and *Is Religion Irrational?* (2011).

---

1. What initially drew you to theorizing about science and religion?

Science and religion share some common questions: what is the origin of the universe? Has it a purpose? What is the human self? Why are the laws of nature as they are? So for me it has always seemed natural to look at the varied answers they give. Since science gives experimentally based answers to these questions, it is vital to take scientific findings into account, if you want to understand the universe. But questions about value, meaning, and purpose are not addressed by science—certainly not by the 'hard' sciences like physics, chemistry, and biology. So there are questions to be asked about how these other questions relate to scientific data. I have always been fascinated by this topic.

2. Do you think science and religion are compatible when it comes to understanding cosmology (the origin of the universe), biology (the origin of life and of the human species), ethics, and/or the human mind (minds, brains, souls, and free will)?

Since the sixteenth century, advances in the sciences have completely changed the human view of the universe. I am in no doubt that well-established scientific conclusions are to be accepted. Any religious views that deny such conclusions must be rejected. If scientific investigations showed that life after death was impossible, or that non-physical entities (like God) could not exist, that would dispense with any such religiously based beliefs. But the hardest thing for many scientists to recognise is that the sciences have their limits. Physical sciences obviously cannot directly deal with non-physical (spiritual) realities. Nor can they deny that such realities exist (though they may claim there is no evidence for them in the physical world). It follows that the hardest compatibility question is whether the existence of spiritual realities looks unlikely or impossible on present scientific information.

In the case of cosmology, to take as an example just one science, it could be claimed that science provides a complete explanatory account of the origin and nature of the cosmos. This, however, is clearly false, since there are untold numbers of things we do not understand about the cosmos, as any competent cosmologist knows. We just do not have enough information about causal processes to decide whether there is a spiritual agency (God) at work.

We could say that we do not need to appeal to spiritual agencies in cosmology, but that is a question of methodology. We simply exclude what lies outside the sorts of explanation we employ. For me, a key principle of science is: each science (and there are many, not all reducible to each other) selects a set of physical data, mathematical techniques, experimental methods, and technical concepts. It does not deny that there are other scientific concepts and methods (physicists do not deny that biology exists), but it does not deal with them. There is no one super-science, not even physics, the favourite candidate. It follows that there may well be non-scientific facts too—facts about ethics, art, human conscious experience, for instance.

How these areas relate to one another is a highly disputed question. It is not about compatibility so much as it is about the coherence or 'fit' of the various departments of human knowledge. So I think the problems of coherence between different scientific conclusions, and between any set of such conclusions and the data of personal experience, are fascinating and complex and unresolved. In my view, we have no convincing 'big picture'. Those who think the natural sciences provide all relevant information seem to be almost certainly wrong, however.

3. Some theorists maintain that science and religion occupy non-overlapping magisteria—i.e., that science and religion each have a legitimate magisterium, or domain of teaching authority, and these two domains do not overlap. Do you agree?

I think the picture of non-overlapping magisteria is misleading. It bundles all the sciences (from physics to psychology and economics) together misleadingly, does the same with all the humanities (of which religion is one aspect), and then juxtaposes these two invented 'blocks' of knowledge. The picture is more diverse, and requires close analysis of specific sciences and specific claims made in specific religions. When you do that, of course, no general answer about the relations of 'science in general' and 'religion in general' can be sensibly given.

Take, for example, the question of whether there is a human soul. Clearly, findings in neuro-science are very relevant to this question, so there is an 'overlap'. But there are many different doctrines of the soul, even within Christianity. Many Christians espouse 'non-reductive materialism', which requires something like 'whole-part' causation within complex physical systems. Can neuro-science come to a decision as to whether there is such a thing? Not at present, it seems, as the issue is unresolved and highly disputed within neuro-science. Also, while many neuro-scientists assume that every entity must have a physical location and nature, it does not seem that their researches establish this. What they can say is that specific parts of the brain must function properly to enable mental functions to operate—i.e. there is a strong correlation between physical processes and conscious states. We need a lot more work to see how to advance in this area. On the whole, religious views would oppose reductive materialism, but we need to explore how much such a materialism is needed in science—I suspect not at all. There is plenty of work to be done.

4. What do you consider to be your own most important contribution(s) to theorizing about science and religion?

I am not a working scientist, but I am trained in philosophy and theology. So my main contributions to this area are in exploring in detail what religious views about the nature of reality are, and what revisions they allow. Because there is a materialistic and anti-religious element to some popularising science writing, I have also tried to expose the fact that this element arises from a philosophical version of materialism which is quite weak, and that the sciences are not committed to. More controversially, I have argued strongly for philosophical idealism—a range of views which claim that mind, or something mind-like, is the basis of all reality, including physical reality. Obviously this will have implications for the physical world—it will presumably exemplify value and purpose in some form—but there are many different ways in which this could be the case. Since my books are mostly read by religious believers, my main purpose in that respect has been to encourage a love of science, and to see the implications of scientific discoveries for religious beliefs.

5. What are the most important open questions, problems, or challenges confronting the relationship between science and religion, and what are the prospects for progress?

I am a Christian theologian, and my main interest is in expounding a wholly reasonable form of Christian belief that is not inconsistent with well-established modern knowledge, including scientific knowledge. From that point of view, the main challenge is to examine the claims of those who argue that science is in some way incompatible with religious faith, or that religion is irrational or not based on evidence. Since there are hundreds of good scientists who are religious believers, I think the prospects for success are good. It will be necessary to re-formulate many ancient Christian beliefs, and it will also be necessary to establish the limits of scientific explanation, and to distinguish experimentally based claims from more general philosophical claims (especially the great intellectual antitheses, materialism and idealism). I am open to the possibility that I am wrong, and that my whole life's work has been based on a mistake. Naturally, however, I hope this is not the case. Looking at the matter as dispassionately as I can, I think it is indisputable that science, awe-inspiring as it is, is still in its infancy and there are vast gaps in our knowledge of the universe. I also doubt that millions of intelligent and morally heroic religious believers could have been deluded or suffered some sorts of illusions. So I think we need to get away from the rhetoric of abuse and stereotypical thought which is so frequent these days. We need to pursue detailed and patient work in understanding the relationship of specific religious and scientific claims in all their diversity. And we need to recognise that none of us can claim certainty when the universe is so much vaster and more complex than we had ever imagined.

33

Rabbi David Wople

David Wolpe is Rabbi of Sinai Temple in Los Angeles, California, author of several books, and a regular contributor to *Time*, the *Jewish Week*, *The Washington Post*, and many other publications. He has taught at the Jewish Theological Seminary of America, The American Jewish University, Hunter College, and UCLA. Rabbi Wolpe was named the "most influential Rabbi in America" by *Newsweek Magazine* (2012), one of the "50 most influential Jews in the world" by *The Jerusalem Post* (2012), and one of the "hundred most influential Jews in the United States" by *Forward* (2003). His books include *The Healer of Shattered Hearts: A Jewish View of God* (1990), *In Speech and In Silence: The Jewish Quest for God* (1992), *Teaching Your Children About God: A Modern Jewish Approach* (1993), *Why Be Jewish?* (1995), *Floating Takes Faith: Ancient Wisdom for a Modern World* (2004), and *Why Faith Matters* (2008).

1. What initially drew you to theorizing about science and religion?

It all began at the dinner table. My father was a Rabbi and a man of wide intellectual interests and even passions. My oldest brother is a natural scientist (a developmental biologist), my second oldest brother a social scientist (a bioethicist), and my younger brother and I became Rabbis. Questions of science, society and religion were the subject of constant, ardent debate. Talmudic questions and rat embryo injections jostled one another for disputational priority. Nothing was ruled out of court because it belonged to this or that discipline. The effectiveness and importance of science were never questioned. Yet its comprehensiveness as an explanation of everything was never taken for granted. My interest in the field of science and religion is a natural—perhaps inevitable—outgrowth of the home in which I was raised.

2. Do you think science and religion are compatible when it comes to understanding cosmology (the origin of the universe), biology (the origin of life and of the human species), ethics, and/or the human mind (minds, brains, souls, and free will)?

Science poses some insurmountable challenges to certain conceptions that exist in certain religions. Only the charlatan and the credulous can maintain that the world was created in a literal seven days or that dinosaurs are a fiction. The amount of intellectual quackery and narrowness that finds a home in religion is dispiriting. It is however, a dilution and not a reflection of its deeper spirit. And we must at least be mindful that science has been hospitable to a fair share of trickery and simple mindedness as well.

Properly understood science is a boon to faith, not a barrier. The marvel that the world yields to our researches, that it is governed by perceptible laws, that its regularity reveals what can only be called majesty—these are reinforcements of the attitude of sanctity and reverence with which religion imbues the world. Approaching religion as a stock of insight, analogy and poetry erases many of the conflicts: The cosmology of the Psalms is a metaphor, not an astronomical hypothesis. We may as well investigate the images of Wordsworth for botanical verisimilitude. Brain research is a difficult and challenging area. (And also perhaps the most fascinating intellectual adventure of our time.) To posit a soul when one knows that much of what we are resides in neuro-wiring and can be changed or damaged, is a genuine puzzlement. If I open your head and mess with the matter encased in your skull, you will emerge very different. Have I altered your 'soul'? Or is such an idea a fiction? There are responses: If we conceive of the body as the receptacle, like a radio, through which a soul moves, then if you damage or alter the hardware the mechanism will not be the same. But for those who wish to preserve the idea of a soul, this is a problem that deserves more thought and creativity.

3. Some theorists maintain that science and religion occupy non-overlapping magisteria—i.e., that science and religion each have a legitimate magisterium, or domain of teaching authority, and these two domains do not overlap. Do you agree?

Yes. And no. There are distinct areas of interest, but I would not say they are thoroughly non-overlapping. Theories of human nature for example, developed by religion and philosophy, are supplemented, corrected and disputed by science. Neuroscience has proved invaluable to our understanding of the way people work, and self-deluded (or terrifically blinkered) is the religious thinker who denies it.

On the other hand, the notion that "is" implies "ought" is no more true in sociological research than in classical moral theory. We may understand our moral mechanisms better through research (though this at least is disputable) but how we should behave remains a preserve that is not biological, but what I would call spiritual and ethical. And one's fundamental orientation toward the world—as well as one's fundamental purpose—are not scientific questions. They cannot be decided in labs or by statistical surveys. Internal experience is ungraspable by science but determinative in life. Meaning cannot be spun in a centrifuge and the most basic questions on which our lives turn are not empirical.

4. What do you consider to be your own most important contribution(s) to theorizing about science and religion?

In my own writing and debates on these questions I have tried to emphasize two related points. First is to dispel the simple-minded notion that pulpit Rabbis and Pastors, religious thinkers outside the academy, do not learn about science or take interest in these questions. On each side a certain smugness has crept into the debate and honest, thoughtful engagement is the best way to persuade people that neither side need be close minded.

The second is to enlarge the conception of what religion is beyond a recitation of dogma to an orientation toward the universe. It is a wonder and an awe that arises in part from the acknowledged limitations of our cognition. After all, assuming we are purely products of evolution, our minds are bounded by the same limitations as our other organs of sight or smell or sense. So we grasp a small slice of the spectrum of all that can be known. Therefore any unqualified assertion about the nature of reality is automatically suspect, and the idea that we can intuit more than we can prove or demonstrate should be given room to breathe. Our sense of the sacred is to the religious an irreducible category, a perception of something real, and part of the basis of a full life.

5. What are the most important open questions, problems, or challenges confronting the relationship between science and religion, and what are the prospects for progress?

From the point of view of religion, the greatest challenge is the scientistic mindset (not the scientific mindset, but the scientistic one). Such an unnuanced view understands science as the only, and only possible, manner to answer the questions of life, and all that does not fall within its purview is automatically rendered irrelevant or obscurantist. So long as that view holds sway among protagonists there can be no mutuality.

Among religions, obscurantism of another kind is the great danger. Science is so powerful that it frightens religions into defensive crouches, denying its potency and cogency. Religion has to open to science, integrate its teaching, modify its excesses, measure its discoveries against the cherished convictions of the believer. The Rabbis of the Talmud teach that "God's seal is truth." Therefore if it is true, it cannot be 'ungodly.' Together, science and religion can contribute to a rich, thick culture in which all can find a place for their minds, their hearts and their spirits.

About the Editor

Gregg D. Caruso is Associate Professor of Philosophy and Chair of the Humanities Department at Corning Community College, SUNY. He is also Editor-in-Chief of the peer-reviewed journal *Science, Religion and Culture* (SRC). He received his B.A. in Philosophy from William Paterson University and his M.Phil and Ph.D. in Philosophy from the City University of New York, Graduate Center. He is the author of *Free Will and Consciousness: A Determinist Account of the Illusion of Free Will* (2012) and the editor of *Exploring the Illusion of Free Will and Moral Responsibility* (2013). Dr. Caruso's research interests include philosophy of mind, cognitive science, and metaphysics, with a particular interest in consciousness and free will. His other interests include science and religion, ethics, social and political philosophy, and moral responsibility. Currently he is Vice President of the Southwestern Philosophical Society (SWPS), contributor to the blog *Flickers of Freedom*, and an Assessing Editor for *The Journal of Mind and Behavior*.

Index

CPSIA information can be obtained at www.ICGtesting.com
Printed in the USA
BVOW03s1328230414

351463BV00002B/192/P